… à Bâillon

Kant

[handwritten annotation]

CIRCULAR.

HEAD-QUARTERS DISTRICT OF COLUMBUS.
6TH DIVISION, 16TH ARMY CORPS,
COLUMBUS, KY., June 28, 1863.

The attention of Company Officers is called to a lately published work, by J. B. Lippincott & Co., of Philadelphia, entitled

" THE COMPANY CLERK: *showing how and when to make out all the Returns, Reports, Rolls, and other Papers, and what to do with them. How to keep all the Books, Records, and Accounts required in the Administration of a Company, Troop, or Battery in the Army of the United States. By Captain August V. Kautz, 6th U. S. Cavalry; Colonel 2d Ohio Vol. Cavalry.*"

This work can be purchased at the bookstores in Columbus and Cairo, at St. Louis, &c.; and all officers in the volunteer service are recommended to acquaint themselves with its very valuable and necessary information.

By order of BRIGADIER-GENERAL ASBOTH.

T. H. HARRIS,
Assistant Adjutant-General.

THE

COMPANY CLERK:

SHOWING

HOW AND WHEN TO MAKE OUT ALL THE RETURNS, REPORTS, ROLLS, AND OTHER PAPERS, AND WHAT TO DO WITH THEM.

HOW TO KEEP ALL THE BOOKS, RECORDS, AND ACCOUNTS REQUIRED
IN THE ADMINISTRATION OF A COMPANY, TROOP, OR BATTERY
IN THE ARMY OF THE UNITED STATES.

BY

CAPT. AUGUST V. KAUTZ,

SIXTH U. S. CAVALRY; COLONEL SECOND OHIO VOL. CAVALRY.

PHILADELPHIA:
J. B. LIPPINCOTT & CO.
1864.

PREFACE.

WE have numerous handbooks for military service that tell us *what* to do, but few, if any, that tell us HOW TO DO IT; and although there is certainly none better than the Regulations itself upon the subject of administration, yet that it essentially fails to accomplish the object, is apparent in the want of care of public property, the informality and want of method in the keeping of the records, and the total neglect, in most of the regiments, to render the prescribed returns. This neglect is believed to be caused more by the difficulty of finding out *how* to do what is required, than from any innate carelessness or intent to neglect their duty on the part of officers.

The want of a book to explain these administrative matters more in detail has induced the author to undertake the task so far as relates to the Company. While he has tried to lay down the practice in the Regular service, he does not presume to teach the older officers of the army their duty; but he trusts they will not be too critical on what is very plain to them, but what is certainly very obscure to the young officer when he first enters service. What has here been prescribed for the Regular service is equally applicable for the Volunteer service; and it is mostly for their benefit that the following pages have been undertaken.

CONTENTS.

LIST OF ABBREVIATIONS USED IN THIS BOOK AND IN OTHER MILITARY WORKS, AND IN MAKING OUT OFFICIAL PAPERS.

A. A. A. G.—Acting Assistant Adjutant-General.

A. A. G.—Assistant Adjutant-General.

A. A. Q. M.—Acting Assistant Quartermaster.

A. C. S.—Acting Commissary of Subsistence.

A. D. C.—Aid-de-camp.

A. G. O.—Adjutant-General's Office.

Act March 3, 1863.—Act of Congress approved March 3, 1863.

Adjt.—Adjutant.

Art.—Artillery.

Art. 35.—Thirty-Fifth Article of War.

Asst.—Assistant.

Bat.—Battalion.

Batry.—Battery.

Brig.—Brigadier, Brigade.

Bug.—Bugler.

Bvt.—Brevet.

C. S.—Commissary of Subsistence.

Capt.—Captain.

Cav.—Cavalry.

Cdt.—Cadet.

Chap.—Chaplain.

Co.—Company.

Col.—Colonel.

Comdg.—Commanding.

Comdt.—Commandant.

Corp.—Corporal.

Dept.—Department.

Det.—Detachment.

Div.—Division.

Drag.—Dragoon.

Eng.—Engineer.

Ens.—Ensign.

Far.—Farrier.

Ft.—Fort.

G. O.—General Order.

Gen.—General.

Hd.-Qrs.—Head-Quarters.

Hosp.—Hospital.

Hosp. Stwd.—Hospital Steward.

Inf.—Infantry.

Inspr.—Inspector.

J. Advt.—Judge-Advocate.

L. Art.—Light Artillery.

Lieut. and *Lt.*—Lieutenant.

M. R.—Mounted Rifles.

M. S.—Medical Staff.

M. S. K.—Military Storekeeper.

Maj.—Major.

Med. Cdt.—Medical Cadet.

Med. Dept.—Medical Department.

N. C. O.—Non-Commissioned Officer.

7

O. B.—Official Business.
Ord.—Ordnance.
Ord. Sergt.—Ordnance Sergeant.

P. D.—Pay Department.
P. M.—Paymaster.
Par.—Paragraph.
Pvt.—Private.

Q. M.—Quartermaster.
Qrs.—Quarters.

R. C. S.—Regimental Commissary of Subsistence.
R. Q. M.—Regimental Quartermaster.
Rct.—Recruit.
Reg.—In this book, *Revised* Regulations of 1861.
Regt.—Regiment.
Regtl.—Regimental.

S. O.—Special Order, Signal Officer.
Sdlr.—Saddler.
Sec.—Section.
Sergt.—Sergeant.

Sergt. Maj.—Sergeant-Major.
Servt.—Servant.
Sub. Dept.—Subsistence Department.
Supt.—Superintendent.
Surg.—Surgeon.
Surg. M.—Surgeon's Mate.

Top. Eng.—Topographical Engineer.
Trptr.—Trumpeter.

U. S. A.—United States Army.
U. S. Art.—United States Artillery.
U. S. Cav.—United States Cavalry.
U. S. Eng.—United States Engineers.
U. S. I.—United States Infantry.
U. S. M. D.—United States Medical Department.
U. S. T. Eng.—United States Topographical Engineers.

Vol.—Volunteers.
Vouch.—Voucher.

W. D.—War Department.

THE

COMPANY CLERK.

1. THE company clerk is a non-commissioned officer or soldier, selected on account of his penmanship and capacity for keeping the records and making the reports and returns required from the company. He is directed and instructed by the commanding officer of the company and the first sergeant. He is reported on daily duty, and does not receive any additional pay, but, whilst employed, is excused from all other duties possible with the company. In times of active service, however, he should always be prepared for service in the field.

The following are the books and records he is required to keep, viz. :

Books.

2. MORNING REPORT BOOK,
SICK BOOK,
ROSTERS,
DESCRIPTIVE BOOK,
CLOTHING BOOK,
ORDER BOOK,
ACCOUNT BOOK OF COMPANY FUND,
REGISTER OF ARTICLES ISSUED TO SOLDIERS,
RECORD BOOK OF TARGET PRACTICE.

Reports, Returns, Rolls, and Papers.

3. The following reports, returns, rolls, and other papers are required to be made out by him, viz.: ·

DAILY.—List of Sick, in the Sick Book.
> Morning Report, in the Morning Report Book
> Details of men for guards, detachments, and fatigue.

MONTHLY.—Monthly Return.

BI-MONTHLY,—viz.: at the end of February, April, June, August, October, and December, viz.:
> Muster Rolls.
> Report of Damaged Arms.

QUARTERLY,—viz.: at the end of March, June, September, and December:—
> Return of Clothing, Camp and Garrison Equipage.
> Return of Ordnance and Ordnance Stores.
> Return of Quartermaster's Property.
> Return of Deceased Soldiers.
> Descriptive list of men joined.
> Return of Blanks.

QUARTO-MONTHLY,—viz.: at the end of April, August, and December:—
> Return of Company Fund.

ANNUALLY.—Annual Return of Casualties.

Papers.

4. In addition to the foregoing papers, the following are also required when circumstances render them necessary.

Certificates of Disability.
Final Statements.
Discharges.
Descriptive Rolls.

Furloughs, Passes, Sick Furloughs, &c.

Affidavits, Certificates, &c.

Inventories of Deceased Soldiers.

Proceedings of Company Council of Administration.

Provision Returns.

Requisitions for Forage, Fuel, Stationery, Straw, and for every kind of Property, as Arms, Accoutrements, Ammunition, Clothing, Camp and Garrison Equipage, Quartermaster's Property, and, in fact, every thing required by a company.

Inventories for Inspection Reports of Property to be inspected and condemned.

Inventories of Damaged Property for Boards of Survey.

Letters of Transmittal, Complaints of Soldiers, Applications for Transfer, &c.

Returns of Killed, Wounded, and Missing in Action.

Reports of Target Practice.

Charges and Specifications.

The foregoing will be treated of in the order in which they are likely to come to the attention of an inexperienced person.

General Observations.

5. It will be remarked that the foregoing books and papers relate to two objects, viz. : the men and the property.

What relates to the men is a record of their accounts and services ; that these should be correct and perfect, is a matter of justice to the men and to the Government.

6. What relates to property is a matter of account between the officer and the Government. The property is intrusted to his charge, he is paid for taking care of it, and required to show how he has done so. To do this, he must remember the following general principles, viz. :

7. To get invoices of all the property he receives.

8. To take receipts for all the property he transfers.

9. To get certificates for property lost or destroyed or expended, from other officers, if possible, or affidavits from soldiers or citizens, or, finally, his own certificate.

10. To bear in mind not to get the property of different departments, as Ordnance, Quartermaster's Property, or Clothing, Camp and Garrison Equipage, mixed on the same paper.

11. Not to allow unserviceable property to accumulate, except to save it, and at the earliest opportunity bring it before an inspecting officer and have it disposed of.

12. If these facts are borne in mind, an officer can always account for the property with which he has been intrusted, even though he is not able to make the returns at the prescribed time, as frequently happens in active service.

Morning Report.

13. The Morning Report of each company should be sent into the adjutant's office every morning after sick call, and before eight o'clock, and should be signed by the first sergeant, or, in his absence, by the acting first sergeant, and by the officer in command of the company, and should be an accurate report of the company for the day.

14. On the page for Remarks should be entered, by name, the changes in the figures from the previous report. To facilitate this, each month should be commenced by entering the name of every officer and soldier "*not for duty*," whether present or absent.

Thus, beginning with the heading "Special duty," *Present,* and bearing in mind that officers only are detailed on such duty: Second Lieutenant B, on special duty, Special Orders No. 5, dated Head-Quarters 2d U.S. Cavalry, Nov. 10, 1862; Sergeant C, and private D, on extra duty; Corporal L, and private M, on daily duty; Captain G, Sergeant N, Bugler F, Saddler K, and private L, sick;

First Lieutenant M, and Corporal L, in arrest; blacksmith P, and private S, in confinement. *Absent.*—Captain A, Sergeant L, and privates M and W, on detached service, per Order No. 10, dated Head-Quarters, Department of Kansas, Nov. 5, 1862; private O, with leave; private F, without leave; private R, sick; private L, in confinement. *Gain.*—Captain A, from detached service; private M, recruit from depot; private L, enlisted; private N, from desertion. *Loss.*—First Lieutenant M, resigned; private F, died; Corporal L, killed in action; private Z, deserted.

On the second and every day for the balance of the month, it is only necessary to mention by name the changes since the last report, thus: Second Lieutenant B, from special duty to duty; private L, from sick to duty; private K, duty to sick; private S, from confinement to sick, &c. Thus the condition and whereabouts of the men of the company may be learned without referring back; and the remarks might be made still more explicit by stating "where" and "since when" they have been absent, and by what authority; also the number and date of order should be given, when there is any that has caused the change, as when men are detached or detailed on extra duty.

15. Every man is for duty unless excused by the surgeon, or on some other duty which relieves them from duty with the company by competent authority. Captains or officers commanding companies have no authority to excuse men from duty except for some legitimate· purpose, as the company clerk, or tailor, or blacksmith.

Sick Book.

16. In order that the report of sick may be correct, each company should have a "Sick Book," for which a small blank-book the size of note paper may be used, and each page ruled out, (see Form 1).

2

Form 1.

Company A, 1st U. S. Cavalry. Names. January 1, 1863.	In hospital.	In quarters.	For duty.	Excused.	REMARKS.
Sergeant, Smith, J. H..........	1	.....	.....	1	
Corporal, Jones, B..............	.....	1	.....	1	Not permitted to leave quarters or camp.
Privates, Brown, H.............	.....	.....	1	.....	
Thomas	.....	.....	1	.....	Except riding and guard.
Jan. 2.					
Sergeant, Smith, J. H..........	.....	1	.....	1	Must remain in quarters.
Corporal, Jones, B..............	.....	.....	1	.....	
Privates, Brown, H.............	1	.....	.....	1	
Thomas	.....	.....	1	.....	Excused from riding.
Johnson...............	1	.....	.....	1	
Jan. 3.					

17. The Sick Book should contain all the names of the sick, or who represent themselves as such, and the surgeon decides whether they are to be excused or not, and no other person has the authority to decide so, with regard to the sick. A non-commissioned officer should always take the Sick Book and march the sick to the surgeon's quarters or office immediately at sick call.

18. The Surgeon enters upon the Sick Book his decision, and the non-commissioned officer takes the book back immediately to the first sergeant, that he may complete his morning report. "For duty" means all those who are fit for the legitimate duties of the soldier, as guard duty, drills, and parades, or detached service.

19. The remarks should specify from what duty the soldier is excused, where he is only partially incapacitated for duty, and they may also contain the surgeon's injunctions with regard to care in quarters, which the first sergeant should have complied with.

Roster.

20. The Roster is a blank-book ruled after the manner of a time book, with the numbers of the month running across the top of the pages, and a column on the left for the names, on which a record of the various details of the company is kept. (See Form 2.) The Rosters are kept for three classes of duty, viz.: 1st. Guard; 2d. Detachments; 3d. Fatigue.

21. The rule which governs in making details is that the longest off are the first for duty. The Regulations say (Par. 566) that when a soldier is sick his tour passes. This, however, cannot mean that it passes as if he had gone on the tour. When he is for duty again, he is manifestly the next for detail. If the detail for the three classes

PAR. 20.

Form 2.

Roster for Guard for Company A, 1st U. S. Cavalry. for the month of January, 1863.

Names.	1	2	3	4	5	6	7	8	9	10	11	12	13	14	15	16	17	18	19	20	21	22	23	24	25	26	27	28	29	30	31	REMARKS.
Arnold..........	1																															
Brown..........	1																															
Camp...........	Sick		1																													
Davis...........	1																															
Edwards	1																															
French		1																														
Laddis		1																														
Harris		1																														
Jones..........		1																														
King...........		1	1																													
Long...........		1	1																													
Murry..........				1																												
Peters..........				1																												
Rogers..........																																
Smith				Sick																												
Thomas				1																												
Williams.......				1																												
Waters..........				a	..	..	..	..	..	..	..	..	..	..	..	..	..	..	..	..	..	..	..	..	..	..	..	..	..	..	..	Absent on detached service.
Yale...........				Ex.	..	..	..	..	..	..	..	..	..	..	..	..	..	..	..	..	..	..	..	..	..	..	..	..	..	..	..	
York...........					..	..	..	..	..	..	..	..	..	..	..	..	..	..	..	..	..	..	..	..	..	..	..	..	..	..	..	Company clerk.

NOTE.—As the Rosters for Non-commissioned Officers for the various duties of Guard Detachment and Fatigue are kept at the Adjutant's office, their names are not entered on these Rosters in the Company should happen on the same day, that for *guard* takes precedence, for detachments next, and fatigue last. If the detail parades and marches on duty, it receives credit for a tour, although dismissed immediately after. If a detail is dismissed on parade, it does not receive credit for a tour. If a detail for detachment marches out of camp or garrison, it receives credit for a tour. The details are made usually in the evening, to march on in the morning, or in the morning to march on in the evening. It is recommended not to credit the tour until the detail has marched on, checking them only when they are entitled to the tour. The credit is given by entering a figure opposite, the name, under the date of the month at the top of the page. (See Reg. 562 to 572.)

22. A separate Roster should be kept for each class. The details are made on small slips of paper, and published to the company usually at Retreat or Reveille. In cavalry and artillery, stable guards form a separate roster, and generally do duty without arms : they are not regarded as sentinels, and do not challenge. On this roster the duty sergeants and corporals also serve, for whom a separate roster is also kept in the company. They are held responsible that the sentinels are properly posted and relieved at the proper time.

23. A roster of passes and furloughs should be kept, in order that these indulgences be equally distributed throughout the company.

Company Clothing Book.

24. The Company Clothing Book is a blank-book furnished by the quartermaster's department, for the purpose of keeping a record of the issues of clothing to the non-commissioned officers and privates of the company. Va-

STATEMENT *of the cost of Clothing, Camp and Garrison Equipage, for the Army of the United States, until further orders, with the allowance of clothing to each soldier during enlistment, and his proportion for each year.*

CLOTHING.	Engineer Troops.	Hospital Stewards.	Ordnance Sergeant.	Ordnance Mechanics.	Cavalry.	Light Artillery.	Artillery.	Infantry.	First.	Second.	Third.	Fourth.	Fifth.	Allowance for five years.
Uniform Hat	$1 68	$1 68	$1 68	$1 68	$1 68	$1 68	$1 68	$1 68	1	1	1	1	1	5
" Feather	12	14	15	3	15	15	15	15	1	1	1	1	1	5
" Cord and tassel	14 20	14 20	14 20	2	14 20	14 20	14 20	14 20	1	1	1	1	1	5
" Eagle														
" Castle	10			5										
" Shell and flame														
" Crossed sabres					3	3	3							
" Crossed cannon														
" Bugle								3						
" Letter	1	1	1	1	1	1	1	1						
" Number	1	1	1	1	1	1	1	1						
Cap (Light Artillery)						1 06								
" Tulip						8								
" Cord and tassel						75								
" Plate						4								
" Rings, pairs of						8								
" Hair plume						73								
Forage Cap	56	56	56	56	56	56	56	56	1	1	1	1	1	5
" Cover	18	18	18	18	18	18	18	18	2	2	2	2	2	8
Uniform Coat, Musicians'	7 45 7 21	7 21	7 21	7 21	5 97 5 55	5 97 5 55	7 45 7 21	7 45 7 21	2	1	1	2	2	8
" Privates'									1	1	1	1	1	8
Jacket, Musicians'					5 97	5 97								
" Privates'					5 55	5 55								
Chevrons, pairs, N.C.S.	35		1 25		35	35	1 25	1 25						
" 1st Sergeants'	24				24	24	35	35						
" Sergeants'	20				20	24	24	24						
" Corporals'					20	20	20	20						

Item	Price								Quantities
Caduceus									
Shoulder Scales, pairs, N.C.S.	50	90	50	50	50	50	50	50	1
" " " Sergeants'	50	50	50	50	50	50	50	50	2
Trowsers, Sergeants'	3 75	3 75	3 75	3 55	3 75	3 75	3 75	3 75	18
" Corporals'	3 75	3 75	3 75		3 75	3 75	3 75	3 75	18
" Privates'	3 65				3 65	3 65	3 55	3 55	13
Sash	1 84	1 84	1 84		1 84	1 84	1 84	1 84	
Flannel Sack Coat (unlined)	2 40	2 40	2 40	2 40	2 40	2 40	2 40	2 40	
" " " (lined)	3 14				3 14	3 14	3 14	3 14	
Knit Jackets	2 70	2 70	2 70	2 70	2 70	2 70	2 70	2 70	16
Flannel Shirts	1 46	1 46	1 46	1 46	1 46	1 46	1 46	1 46	3
Knit "	1 30	1 30	1 30	1 30	1 30	1 30	1 30	1 30	11
Flannel Drawers	95	95	95	95	95	95	95	1 00	3
Knit "	1 00	1 00	1 00	1 00	1 00	1 00	1 00	1 00	20
Stockings	32	32	32	32	32	32	32	32	4
Bootees, sewed	2 05	2 05	2 05	2 05	2 05	2 05	2 05	2 05	20
" pegged	1 48	1 48	1 48	1 48	1 48	1 48	1 48	1 48	4
Boots, sewed					3 25	3 25			
" pegged					2 87	2 87			
Great Coats	9 50	9 50	9 50	9 50	11 50	11 50	9 50	9 50	1
Straps, pairs	14	14	14	14	50	50	14	14	1
Blankets, woolen	3 60	3 60	3 60	3 60	3 60	3 60	3 60	3 60	2
" painted	1 65	1 65	1 65	1 65	1 65	1 65	1 65	1 65	
" rubber	2 55	2 55	2 55	2 55	2 55	2 55	2 55	2 55	
Poncho, painted					2 10	2 10			
" rubber					2 90	2 90			
Leather Stocks	10	10	10	10	10	10	10	10	
Legging, leather									1
" linen						1 25			1
Overalls	1 58				75	75			
Stable Frocks					1 65	1 65			5
Talmas					5 00	5 00			2

* Mounted men may, at their option, receive *one* pair of "boots" and *two* pairs of bootees, instead of *four* pairs of bootees.

Metallic eagles, castles, shell and flame, crossed sabres, crossed cannon, bugles, letters, numbers, tulips, plates, shoulder-scales, ring, the cap cord and tassels, the hair plume of the light artillery, the sashes, knapsacks and straps, haversacks, canteens, straps of all kinds, and the talmas, will not be issued to the soldiers, but will be borne on the return as company property while fit for service. They will be charged on the muster-rolls against the person in whose use they were when lost or destroyed by his fault.

Camp and Garrison Equipage.

Knapsacks and straps	$2 14	Drum sticks, carriage	$0 52
Haversacks, unpainted	48	" cord	30
" enamelled and painted	56	" snares, sets	16
Canteen, complete	44	" case	38
" strap leather	15	Wall tent	$35 00
Bedsacks, single	3 00	" " fly	17 00
" double	3 15	" " poles, sets	87
Mosquito bars	3 15	" " pins, sets	39
Axe	83		53 26
" helve	12	Sibley tent	60 00
" sling	53	" " pole and tripod	3 40
Hatchet	32	" " pins, sets	31
" helve	3		63 71
" sling	35	" " stove	2 62
Spade	70	Hospital tent	87 54
Shovel	65	" " fly	33 20
Pickaxe	67	" " poles, sets	2 00
" helve	11	" " pins, sets	1 00
Camp kettle	55		123 74
Mess pan	23	Common tent	21 50
Iron pot	1 15	" " poles, sets	70
Garrison flag	43 00	" " pins, sets	25
" " halliard	3 25		22 45
Storm flag	17 00	Shelter tent, complete	3 25
Recruiting flag	6 50	Tent pins, hospital, large	3
" " halliard	1 00	" wall, "	2
Guidon	12 00	" common, small	1
Camp color	2 28	Regimental book, order	1 35
Standard, for mounted regiments	30 00	" " letter	1 35
National color, artillery and infantry	42 00	" " index	1 46
		" " desc'ptive	2 10
Regimental color, artillery and infantry	63 00		6 26
Color belt and sling	4 50	Post book, morning report	50
Trumpet, with extra mouth-piece	3 37	" guard report	86
Bugle, " " "	3 00	" order	50
Cords and tassels for trumpets or bugles	90	" letter	50
			2 36
Fife, "B" or "C"	50	Company book, clothing	2 00
Drum, complete	5 50	" " descriptive	1 38
" head, batter	75	" " order	52
" " snare	28	" " morn. rep't	1 50
" sling	40		5 40
" sticks, pairs	22	Regimental book, general order	1 30
		Record book, for target practice	56

TABLE *specifying the money value of Clothing allowed to the Army of the United States.*

	NON-COM. STAFF		CHIEF MUSICIAN		FIRST SERGEANT			SERGEANT			
	Cavalry and Lt. Artillery.	Artillery and Infantry.	Cavalry or Lt. Artillery.	Artillery or Infantry.	Engineers.	Cavalry or Lt. Artillery.	Artillery or Infantry.	Engineers.	Ordnance.	Cavalry or Lt. Artillery.	Artillery or Infantry.
1st year......	$64 21	$60 77	$65 05	$61 25	$60 55	$62 45	$58 97	$60 33	$60 77	$62 23	$58 75
2d year.......	34 69	34 27	35 11	34 51	34 95	33 81	33 37	34 84	34 27	33 70	33 26
3d year......	51 62	51 18	52 46	52 66	49 96	49 86	48 38	49 74	50 18	49 64	48 16
4th year......	34 69	34 27	35 11	34 51	34 95	33 81	33 37	34 84	34 27	33 70	33 26
5th year......	46 27	46 48	47 11	46 96	46 26	44 51	44 68	46 04	46 48	44 29	44 46
	229 48	226 97	234 84	229 89	226 67	224 44	218 77	225 79	225 97	223 56	217 89

	HOSPITAL STEWARD.	CORPORAL.			MUSICIAN.			ARTIFICER AND PRIVATE.			
		Engineers.	Cavalry or Lt. Artillery.	Artillery or Infantry.	Engineers.	Cavalry or Lt. Artillery.	Artillery or Infantry.	Engineers.	Ordnance.	Cavalry or Lt. Artillery.	Artillery or Infantry.
1st year......	$60 07	$60 15	$62 15	$58 66	$58 15	$61 99	$58 15	$59 25	$59 25	$61 15	$57 07
2d year.......	33 92	34 80	33 66	33 22	32 86	33 48	32 86	34 20	34 20	33 06	32 62
3d year......	49 48	49 66	49 56	48 08	47 56	49 40	47 56	48 66	48 66	48 56	47 08
4th year......	33 92	34 80	33 66	33 22	32 86	33 48	32 86	34 20	34 20	33 06	32 62
5th year......	45 78	45 96	34 21	44 38	43 86	44 05	43 86	44 96	44 96	43 21	43 38
	223 17	225 37	213 24	217 56	215 29	222 40	215 29	221 27	221 27	219 04	213 37

The allowance to Volunteer troops is at the rate of $42 per annum.

rious methods of keeping this account are practised. That shown below is regarded as the best.

25. The names of the non-commissioned officers and privates are entered in the book in alphabetical order, a number of blank leaves being left after each letter according to the probability of acquisition to the company. A page is devoted to each soldier. The issues are made periodically, by first ascertaining the wants of each soldier, and then making a requisition for the clothing, which is approved by the commanding officer, and the articles are then issued by the quartermaster to the company commander, who issues them to the men, upon a Receipt Roll. (See par. 103.)

26. The articles are entered opposite to the soldier's name, on the receipt-roll. He receipts the same, and the issue and signature are witnessed by some officer of the company, not responsible for the issue: if there is no officer, then a non-commissioned officer. (Reg. 1159.) The total money value of each issue is then computed for each soldier, and entered on his page in the Clothing Book, with the date of the issue. (Reg. 1160.)

27. Care must be taken not to over-issue the yearly allowance, for the reason that "extra issues" should be made on separate receipt-rolls (see note to Receipt Roll), and the amounts are charged on the next subsequent muster-roll, and deducted from his pay, on the Pay Roll, by the paymaster. (Reg. 1155.) These issues are entered on the Clothing Book as "Extra," with a memorandum of the muster-roll on which they have been paid. In computing the soldier's clothing for his descriptive roll, the extra issues which have been paid for are omitted; those which have not been paid for alone are entered.

28. To compute a soldier's clothing, you take the sum of the regular issues and the extra issues which have not been paid for, and the total is the amount to be compared with his allowance, which is obtained by computing the time the soldier has been in service and the allowance per month. In the regular service this is obtained from the last list of prices, published periodically in General Orders from the War Department. (Reg. 1157.) In the volunteer service it is fixed by law at $3.50 per month. (See statement of the cost of clothing, camp, and garrison equipage, General Order, No. 202, Dec. 9, 1862, appended.)

29. If the soldier has drawn more than his allowance, the difference is entered on his final statement, "due the U.S. for clothing overdrawn:" if less, it is entered, "due the soldier for clothing not drawn."

30. A question arises where a soldier incurs a disability very shortly after his enlistment. On his first entrance into service he draws nearly a year's allowance of clothing; if at the end of a month he incurs a disability and is discharged, his allowance of clothing and pay for that period would still leave him in debt to the Government. If the man is unfortunate and the disability the result of no indiscretion on his part, or if he should be entitled to a pension under the law, it would be but just to allow his clothing account to balance, and give him his pay for the period he has been in service. If, on the contrary, the man is of unworthy character, and his discharge the result of his own misconduct, or the result of a fraud, as in the case of a minor, or where the man has brought the disability with him into service, he should be charged the full amount of clothing drawn.

31. Clothing is not charged as "extra" until the soldier has exceeded his yearly allowance, although some officers have followed the practice of settling the clothing account quarterly.

32. Careless and improvident soldiers frequently resort

to selling their clothing for the purpose of obtaining money; and in some localities, where clothing is in demand, they can sometimes obtain a decided advance upon Government prices: under such circumstances great care and frequent inspections must be resorted to to prevent it, and the soldiers be required to produce their clothing and account for its absence.

33. The act of Jan. 11, 1812, prohibits the purchase of soldier's clothing by any person not subject to the Rules and Articles of War, under a penalty of three hundred dollars fine and one year's imprisonment. The act of March 3, 1863, (sec 23,) authorizes any officer, civil or military, to seize and take the "clothes, arms, military outfits, and accoutrements," furnished to soldiers by the United States, in the possession of persons not soldiers, who have obtained them contrary to law or regulation.

34. Soldiers, under the 38th Article, of War, are liable to stoppage of pay, and punishment, for selling clothing or neglecting to take care of it.

Order Book.

35. In this book are entered all general orders from the regimental head-quarters, and also all special orders that affect the company or any member of the company. At orderly call the company clerk repairs, with the orderly sergeant and the Order Book, to the regimental head-quarters, and enters all orders that are to be copied into the Order Book. (Reg. 443.) This book may be divided into parts for each kind of orders, or they may be kept continuously according to date. The verbal orders should also be entered as memoranda, to render their remembrance more certain.

Company-Fund Account Book.

36. This book is kept according to Form 3. The receipts are entered on the left-hand page and the expenditures are entered on the right. Every two months, or oftener, the company commander shall convene the officers of the company, forming a company council for the purpose of appropriating the money. In case of a tie vote, the post commander decides. (Par. 18. Reg. 205.)

37. The money is disbursed by the company commander according to the resolves of the council. The proceedings are entered in the book and signed by all the members of the council. This book is always open to the inspection of the post commander, or, in the field, to the regimental commander. (Reg. 206.)

38. The money should be appropriated only for the benefit of the enlisted men, and is usually expended in procuring such necessaries as are not furnished by any of the departments of the army—such as condiments and vegetables for the messes, stencil-plates for marking clothes and accoutrements, tools for the mechanics in the company for the manufacture of company property, books for a company library, and, in fact, any thing for the benefit of the company that will conduce to the comfort, health, and convenience of the enlisted men of the company.

39. The Company Fund accrues from the sale of the savings of the company rations, and from the distribution pro rata of the post or regimental funds, and also, when located at a post or garrison where the company can cultivate a garden, from such sales of surplus produce as they may have. The orderly or commissary sergeant keeps an account with the regimental or post commissary of all the rations or parts of rations that he does not draw, and at the end of the month the amount is computed at the cost value and the amount

Form 3.

PAR. 40.

Captain A—— B——, 1st U. S. Cavalry, in account current with the Company Fund, Company A, 1st U. S. Cavalry, for January, February, March, and April, 1863.

Dr.

Date.	From what source received.	$	Cts.
1863.			
Jan. 1......	On hand per last account....	150	00
" 30......	From the Company for Company savings....	60	00
Feb. 28......	" " " "	50	00
Mar. 31......	" " " "	62	00
Apr. 30......	" " " "	58	00
" 30......	From Regimental Fund....	100	00
	Total received....	480	00
	" expended....	156	00
	Balance carried to next account....	324	00

Cr.

Date.	How expended.	$	Cts.
Jan. 1......	For vegetables, per bill....	25	00
" 10......	" four brooms, 25 cents each....	1	00
" 20......	" one set of carpenter's tools....	10	00
" 30......	" condiments, as per bill....	15	00
Feb. 20......	" books, as per bill....	30	00
Apr. 10......	" mess furniture, as per bill....	75	00
Apr. 30......			
	Total expended....	156	00

I certify that the above account is correct and just.

A—— B——
Captain Co. A, 1st U. S. Cavalry.

Approved,

J—— D——
Colonel 1st U. S. Cavalry.

in money is paid by the commissary. If there is no money to pay the account, the commissary gives a certified account of the stores, which may be paid by any commissary. (Par. 186.)

Return of Company Fund.

40. The account is made up whenever a new officer takes command, whenever the company is detached from the post or regiment for any permanent period, and every four months, viz. : at the end of April, August, and December (Reg. 206). It is made out according to Form 3, except that a certificate of its correctness is appended. It is made in duplicate, one in the book, and the other on a sheet of letter paper, which are handed in to the commanding officer of the regiment or post for his approval; the latter is retained by him to be forwarded to the department or brigade commander. The expenditures should be supported by receipts, though they go no further than the post or regimental commander, and are returned with the book to the company commander.

Register of Public Property issued to Soldiers.

41. This Register is kept according to Forms 4 and 5, for the purpose of keeping an account of the arms and accoutrements issued to soldiers. This is to keep a check upon the losses, and to enable the captain to know to whom and for what he shall make the charges on the Muster Roll The names of the soldiers are entered on the left, and the articles are charged under their respective headings. When the articles are numbered or lettered, the number or letter should be under the heading, to prevent exchanges or pilfering of accoutrements or arms.

PAR. 41.

FORM 4.

Register of Arms and Accoutrements issued to Company A, 1st U. S. Infantry.

Date	Names	Muskets	Bayonets	Swords and belts	Bayonet-scabbards	Cap-boxes and picks	Cartridge-boxes	Cartridge-box plates	Cartridge-box belts	Cartridge-box belt-plates	Gun-slings	Waist-belts	Waist-belt-plates	Ball-screws	Screw-drivers	Spring-vices	Tompions	Wipers	Remarks
1863. Jan. 1...	1st Sergeant Allison...	1	1	1	1	1	1	1	1	1	1	1	1	1	1	1	1	1	
	Sergeant Tilton.........	1	1	1	1	1	1	1	1	1	1	1	1	1	1	1	1	1	
	Corporal Milton.........	1	1		1	1	1	1	1	1	1	1	1	1	1	1	1	1	
	Musician Knowles......			1															
	Private Arnold.........	1	1		1	1	1	1	1	1	1	1	1		1		1	1	
	" Brown	1	1		1	1	1	1	1	1	1	1	1		1		1	1	

PAR. 41.

Form 5.

Register of Camp and Garrison Equipage issued to Company A, 1st U.S. Infantry.

Date.	Names.	Bugles.	Letters.	Numbers.	Scales.	Knapsacks.	Knapsack-straps.	Haversacks.	Canteens.	Canteen-straps.	Greatcoat-straps.	Spades.	Axes.	Hatchets.	Camp-kettles.	Mess-pans.
1863. Jan. 1...	Sergeant Allison...	1	1	1	1	1	1	1	1	1	1	2	2	2	2	5
	Corporal Milton...	1	1	1	1	1	1	1	1	1	1					
	Private Arnold......	1	1	1	1	1	1	1	1	1	1					

NOTE.—Sergeants or corporals in charge of tents or messes are charged with the spades, axes, hatchets, camp-kettles, and mess-pans issued to their squads or messes. A separate register should be kept for those articles not issued to the soldiers, as for sashes, fifes, drums, bugles, &c. &c.

42. To prevent mistakes, whenever the soldier turns in any of the articles charged against him, they should be cancelled on the Register. When he loses any article, if the officer thinks the circumstances are such as to exonerate him from the payment, he should require the soldier to subscribe to an affidavit (Form Par. 172), which becomes the officer's voucher for dropping the articles from his return. Great care should be taken by company commanders in this matter, as the certainty of having to pay for them renders the soldier more careful of his accoutrements, and leads comrades to refrain from pilfering and to look well after each other's property. A stolen article should, as a rule, be charged against the soldier, unless he can show that the article was taken from him at a time when he was not able to take care of it, in consequence of having duties to perform that required his attention elsewhere. An article lost without being able to tell where, or how, or when, should be paid for also. "Whenever a soldier, by reason of severe illness or wounds, is obliged to be sent to a regimental or general hospital, it is the duty of the commanding officer, he being responsible for the arms of the man, to see that all arms and accoutrements he has in his possession are returned to the company armory. *After an engagement, such officers should make it one of their first duties to see that all Ordnance Property in the hands of such of the men as have been killed or wounded has been gathered up and properly secured.*" (Par. 69. Instructions for making Ordnance Returns.)

Descriptive Book.

43. The Company Descriptive Book is a blank-book furnished by the Quartermaster's Department, in which are entered, in alphabetical order, the names of every non-commissioned officer and enlisted man belonging to, or who may subsequently join, the company. Opposite each name are entered the age, height, complexion, color of eyes and hair, where born, occupation, when, where, by whom, and for what period enlisted. In the column headed "Remarks" are entered such remarks as have a bearing on the soldier's merits, character, and services, such as promotions and appointments, reductions, compliments or medals for distinguished services, punishments inflicted by court-martials, actions in which engaged, and wounds received, and, in fact, every important item to the credit or discredit of a soldier, to enable officers succeeding to the company to have a correct history of the men.

44. The names should be entered on the line next below the red line, devoting each space between the red lines to a name. After each letter one or more blank pages should be left for acquisitions to the company, according to the prevalence of names beginning with the letter: the proportion may generally be obtained from the company itself, as the acquisitions are likely to continue in the same proportion.

45. The page for commissioned officers should be divided up proportionately for the captains, first lieutenants, and second lieutenants separately, devoting about eight lines to the captains, about twelve to the first lieutenants, and the balance to the second lieutenants, as there are likely to be more alterations in the lieutenants than in the captains. Under the heading of "Remarks" should be entered number and date of order of assignment, date of joining, number and date of order relieving the officer from duty with the company, and date of departure.

46. One of the pages for non-commissioned officers should be devoted to sergeants and the other to corporals, with remarks similar to those of the officers.

47. One page of the register of men transferred should be devoted to those who join the company, and the other to those who are transferred from the company.

48. The register of discharges, deaths, and desertions is kept in the order in which they occur. The Remarks against the deserters should state charges for public property which they carry off with them, or for which they are indebted to the Government.

The Descriptive Book should be overhauled monthly, or oftener, and kept corrected to date, and especially after each muster.

Record Book of Target Practice.

49. This book, furnished by the quartermaster (see Target Practice, p. 43), is for the purpose of keeping a record of the target practice of the company. After the 1st of January each year, the exercises, as laid down by the system of target practice prescribed by the Secretary of War, commence, and continue, when the weather permits, until the course is completed. A record of the firing is kept according to the forms laid down in the Target Tactics. The company is divided into three classes, according to the relative merits of firings. After having fired four rounds each, at the respective distances between 150 and 400 yards, the company is reformed according to the number of hits, and again after firing the whole number of prescribed distances. This enables the officer to know the good from the bad marksmen. The number of hits are recorded opposite each man's name at each firing, and summed up, and the soldier who has the greatest number of hits is entitled to wear the company prize, viz., a brass "stadia," and also becomes a competitor for the regimental prize, a silver "stadia," which is worn by the soldier who makes the shortest string, who in turn becomes a competitor for the army prize, a silver medal, on which are inscribed the name, company, and regiment of the soldier who has made the shortest string. These prizes are given once each year. These prizes, except the army prize, are ordnance property, and are worn by the successful marksman until he is excelled by some other marksman at the annual exercises. The Record Book is sent in to the office of the commanding officer of the post or regiment, for his inspection, at the end of each week, or oftener, as he may direct.

Descriptive Roll.

50. When a soldier is detached from his proper company, he is accompanied by his descriptive-roll, which contains a copy of the company records concerning him, and an account of his clothing. If a number are detached at the same time to the same destination, they can all be entered on one roll or on several. The descriptive-roll should not be given into the hands of the soldier, but to an officer or non-commissioned officer, if there is one; if not, it can be sent by mail, which is generally the case, especially when soldiers are ordered away suddenly before their papers can be made out.

51. Blank descriptive-rolls are furnished by the Adjutant-General's Department, and have printed notes on them giving full directions how to fill them out. The following are the notes, viz.:—

Note 1.—The amount of additional pay, if any, under the act of August 4, 1854, must be carefully noted in the exact words used on the Muster Roll.

Note 2.—Likewise, the amount due the soldier for a certificate of merit, or in lieu of a commission, under sec. 4, act of August 4, 1854, in the exact words used on the Muster Roll.

Note 3.—So, also, of any other extra pay for which he may be mustered, ex. gr. as acting hospital steward, as saddler, &c., and which may be still due him.

Note 4.—In the column headed "Bounty paid," must be entered the whole amount hitherto paid him; in that of "Bounty due," the whole amount yet due, on account of the bounty provided by sec. 3, act of June 17, 1850.

Note 5.—The amount of retained pay due, at date, will be carefully stated.

Note 6.—Stoppages for loss or damage done to arms, or other public property, must be noted, and the articles, and particular damage to each, specified.

Note 7.—When stoppages are due, under sentence of a court-martial, a transcript of the same must be entered here; and the amount already stopped must be carefully stated.

Note 8.—In every case of desertion, the date, and that of delivery, or apprehension, must be given, together with a correct transcript of the order of sentence, or pardon.

Note 9.—A correct transcript of the man's clothing account must be made to date, and the amount due to or from him must be precisely stated.

Note 10.—Should the man have been engaged in any action, or skirmish, it must be mentioned, together with time and place.

Note 11.—A full and particular mention will be made of any wounds he may have received in action, or other injury, whilst in the line of his duty.

For further explanation of the notes, see paragraphs 75 to 79 inclusive.

52. All the issues, charges, and incidents that occur to the soldier during his absence are successively entered on the roll by the officers over him, and signed by them. Should he be further detached, the roll accompanies him until he returns to his company, when the items that have been entered during his absence are added to the company records as a part of the soldier's history.

53. "*First.* Descriptive lists and accounts of the pay, clothing, &c. of soldiers will never, where it can be avoided, be given into their own hands. Such papers should be intrusted only to the officer or non-commissioned officer in charge of the party with which they are."

"*Second*. Except in such cases as that of an ordnance-sergeant, specially assigned to duty at a post where there are no troops, and where he cannot be regularly mustered, *no soldier must be paid on a mere descriptive list and account of pay and clothing*, but only on the muster and pay roll of his company, detachment, or party, or on that of a general hospital, if he be there sick or on duty. *No payments will, therefore, be paid to enlisted men on furlough.*" (General Order No. 86, 1862.)

Monthly Return.

54. The company monthly return is made out on printed blanks furnished by the Adjutant-General's Department, and is forwarded on the first day of the month to regimental head-quarters. A copy is retained in the company, in case of the miscarriage of that for the regiment, and to enable the subsequent return to be made out correctly. Great care and uniformity should be observed in order that the return may be copied *verbatim* on the regimental return. The printed notes on the return, which are very concise, must be studied in order to understand them fully.

55. In order to make out the monthly return correctly, it is necessary to have the last return, the morning report-book, and all the orders that have produced changes affecting the officers or men in the company.

56. The following notes are printed on each return:—

Note 1.—*Actions* in which the company, or any portion of it, has been engaged; *scouts, marches, changes of stations, &c.; every thing of interest* relating to the *discipline, efficiency,* or *service* of the company, will be *minutely* and carefully noted, with *date, place, distance marched, &c. &c.*

Note 2.—The name and rank of the officers and soldiers *killed* or *wounded* in action, with *date* and *place*, will be accurately noted.

Note 3.—One copy of this return will be transmitted, on the first of each month, to the adjutant at regimental head-quarters. Blanks will be supplied to companies from the Adjutant-General's office, and their receipt must be promptly acknowledged.

Note 4.—The *date* (with No., date, &c. of order) at which an officer is *assigned* or *transferred* to, *joins* or *rejoins* the company, *assumes* or is *relieved* from the command of it, or from any *special* duty, will be stated against his name; against that of *absent* officers, the *No.* and *date of order*, the *reasons* for and *commencement* of absence, and *period* assigned for same (to be repeated on every return while it lasts).

Note 5.—After the list of absent officers will follow the record of those *resigned, died, &c.* or transferred *from* the company, with No., date, &c. of order, *date, place,* and, in case of death, its *cause.*

Note 6.—The *date, &c.* of all *transfers* to or from the company (with *No.* and *date of order*), of all *apprehensions, discharges, furloughs, deaths, desertions, &c.* will be accurately noted; also the *places* of discharge, death, desertion, &c.

57. *Note* 1.—This note requires a history of the operations of the company in the mouth. There is no prescribed place or form for doing this. It should be entered last, and in some place left blank on the return after every other part of the return is completed: the space under ABSENT, on the back of the return, is the most likely to give room for what is required by this note. A clear and concise statement is all that is required.

58. *Note* 2.—The list of killed and wounded, required by this note, would follow immediately after the record required by Note 1, as there is no place indicated for it on the return. The best place for it is after what is required by Note 6, if there is room, as the spaces correspond to what is required. This list should be entered separately under the respective headings KILLED and WOUNDED. (See par. 251, 252.)

59. *Note* 3.—Is sufficiently manifest. The envelop containing the return should always be addressed to the adjutant of the regiment, when sent by mail, accompanied with a letter of transmittal signed by the company commander and similarly addressed.

60. *Note* 4.—No remarks are necessary opposite the name of an officer *present*, unless he has joined within the month, and then it is not reported in the next; but if he is on some duty besides his company duty, as acting assistant quartermaster, or acting commissary of subsistence, or acting adjutant, then it is reported every month while it lasts. The remark against an absent officer is reported every month while the absence lasts, or until the character of the absence changes: thus, "With leave per special order, (No. 200,) dated War Department, Nov. 10, 1862, for the benefit of his health, since Nov. 15, 1862, for two months." When the two months have elapsed, thus, at the end of Jan. 1863, if no other order is received, he is reported "Absent without leave since Jan. 15, 1863," which is thus reported while such absence continues.

Note 5.—After the list of ABSENT, and in the same column, under the proper headings, RESIGNED, DIED, and TRANSFERRED, the names of such officers are entered. Under the heading "transferred" are entered only those who are transferred *from* the company. If they are transferred *to* the company, they are entered under the appropriate heading according as they are PRESENT or ABSENT, with the proper remark when and by what order. In case of death, the cause is stated in addition to the date and place.

61. *Note* 6.—This note refers only to the non-commissioned officers and privates. They should be entered, by name, in the order in which they occur on the face of the return: thus, first under the heading GAIN should be entered RECRUITS FROM DEPOT, next ENLISTED IN THE REGIMENT,

4

and so on, from left to right, to LOSS, under which come, first, DISCHARGED, with the remark by what authority, as "Expiration of service," "Certificate of Disability," &c., then TRANSFERRED, DIED, MISSING, DESERTED. It will be observed that there are some blank spaces. These are intended for unusual cases, that cannot be placed under any of the headings given. These should also be entered by name on the back of the return, the same as above, under appropriate headings.

62. EXTRA AND DAILY DUTY MEN are also entered under the separate headings EXTRA DUTY, DAILY DUTY, bearing in mind that *extra duty* means those men who are employed in the quartermaster, commissary, or hospital department, and receive additional pay, whilst *daily duty* means those men who are excused from the common duties of the soldier, as guard, police, fatigue, &c. : such are cooks, clerks, tailors, standing orderlies,—remembering that where quartermaster and commissary sergeants, musicians, artificers, farriers, blacksmiths, saddlers, and wagoners, are a part of the organization, though they are not doing the duties of the soldier as above, neither are they on daily duty, and they are not to be reported as such in this column.

63. The ABSENT enlisted men are reported by name in the column under the heading, with the *nature, commencement, period, and place* of absence: as, "Sergt. A. B. Jones, on furlough since Jan. 15, 1863, for twenty days, in Cincinnati, Ohio." "Private J. Smith, sick in General Hospital, since Jan. 20, 1863, at St. Louis, Mo." The columns for both the extra and daily duty men and the absent are intended to give the rank, in order that the number of sergeants, corporals, musicians, artificers, and privates may be summed up in figures.

64. As there are no ruled lines, for the sake of neatness and in order that the space may be economized, both on the face and back of the return, when the names to be entered are numerous, lines in pencil should be drawn, sufficiently apart to be distinct, whilst a greater number of names can be given by entering the number of each rank and their names continuously on the same line for each of the headings *Extra* and *Daily Duty* and *Absent.*

65. The same contractions that are used in the Muster Roll in case of necessity are allowable here. (Par. 85.)

66. When a company is detached, and so remote from the head-quarters of the regiment that this return cannot reach them in less than ten days, a copy is sent (to the head-quarters of the army in time of peace, and) to the Adjutant-General, in order that they may be entered on the regimental return, which must be sent on in blank.

67. The heading of the company returns should always be made out in the name of the captain, whether present or absent.

Muster Rolls.

68. The. MUSTER ROLLS are much the most important papers to make out, as well as the most difficult to execute. Companies are frequently kept without their pay for months, owing to the inability of the officers to make out the Muster and Pay Rolls.

69. On the last day of February, April, June, August, October, and December, these rolls should be prepared. One Muster Roll for the Adjutant-General and three Muster and Pay Rolls are required: two of the latter are for the paymaster, and one for the company, to be retained as a record.

70. To insure despatch and save time, the rolls may be commenced several days before muster, and completed to the column of Present, also all the *Remarks that it is known* must go on the roll, whether the man to whom they relate is present or absent. On the day of muster, the names of those who will be paraded and inspected by the mustering officer may be entered before the muster, and the lines for those who it is known will be absent are filled up with a black horizontal line where the name should be. The doubtful must be left until the muster, when, or immediately after, the roll must be entirely completed, with all the remarks, in order that the mustering officer may send them at once to the Adjutant-General. In case the commanding officer is the mustering officer, he may, if he sees fit, have the rolls completed after *muster*; but it is just as easy to do it before, and far more satisfactory to all parties.

71. The heading of the Muster Roll is made out in the names of the captain and colonel of the regiment, whether they are present or absent. The names of all the officers and men of the company, present or absent, that belonged to it at the last muster, are entered in the first column; the officers and non-commissioned officers, artificers, &c., according to rank, and the privates in alphabetical order. It adds neatness to the roll if each grade is headed, as *Officers, Sergeants, Corporals, Privates, &c.*

72. Those who have ceased to belong to the company *since* the last muster are entered at the foot of the roll, under the headings DISCHARGED, TRANSFERRED, DIED, and DESERTED, respectively, the headings succeeding each other in the foregoing order.

73. The rank, enlistment, and last payment are entered in the succeeding columns, copied from the last Muster Roll and accurately compared.

74. Before entering the remarks, the NOTES on the Muster Roll should be attentively read and studied, as they are very concise, and intended to give all the instructions necessary; but there are so many rolls spoiled by new beginners, as to prove them to be quite incomprehensible to inexperienced persons.

Note 1.—All officers and soldiers are to be taken up on the rolls so soon as *assigned* to the company by *competent authority*, whether they have yet *joined* or not, and to be *dropped* when similarly transferred from it.

Note 2.—Under the head of REMARKS, the *date* when any assignment *takes effect*, the *No., date, &c.* of *order* therefor; the *date* of any officer or soldier's *joining*, whether *originally* or from *any absence*; the *date* of an officer's *assuming* or being *relieved* from any *command*, or *special duty*; the *description* of any *special, extra,* or *daily* duty, on which officers or soldiers may be; all changes of rank, by *promotion, appointment,* or *reduction,* with *date of* same, and *No., date, &c.* of order; all *authorized stoppages, fines, sentences,* with *No., date, &c.* of order, &c.; in case of ABSENCE, the *nature* and *commencement* of, *No., date, &c.* of order, and *period* assigned for same (to be *repeated on every roll while it lasts*); if *wounded* in battle, or *injured on duty,*—if *sick* or *confined,* a remark to that effect; *&c. &c.*—must be *carefully stated,* opposite to the name of the person concerned, *with every thing else necessary, either to account fully for every individual of the company,—to guide the paymaster,—or insure justice to the soldier, and to the United States.*

Note 3.—In noting STOPPAGES to be made for *loss* or *damage* to public property, the *gross* amount due for *Ordnance,* for *Horse equipments,* for *Clothing,* &c., will be *separately* stated, in the order enumerated in par. 1332, G. R.

Note 4.—Additional pay, due under *Sec. 2, Act of Aug. 4, 1854,* will be thus noted, viz.: "*For 1st re-enlist. $2 pr. mo. ;*" or, "*For 2d re-enlist. $3 pr. mo. ;*" or, "*For 3d re-enlist. $4 pr. mo.,*" &c. &c. That due under *Sec. 3 of the same act,* thus: "*For cert. of merit, $2 pr. mo.*" That due under *Sec. 4 of the same act,* thus: "*In lieu of comm., $2 pr. mo.*"

Note 5.—The *instalments* of Bounty due under *Sec. 3, Act of June 17, 1850,* are paid as follows: $\frac{1}{10}, \frac{1}{6}, \frac{1}{6}, \frac{1}{4}$, at the end of the 1st, 2d, 3d, and 4th years respectively, the remainder at the expiration of enlistment: and will, under the head of REMARKS, be noted thus: "*Ret'd Bounty due 1st (or 2d, 3d, &c.) inst. $——*" See G. O. 20 of 1850. *Besides* which, in the columns headed "BOUNTY PAID" and "BOUNTY DUE" must be entered, in figures, the *whole amount hitherto paid,* and the *whole amount yet due* on account of said bounty.

Note 6.—The "three months' extra pay" for re-enlistment under Sec. 29, Act of July 5, 1838, being paid by the recruiting officer, should not be noted on the Muster Rolls.

4*

Note 7.—The roll of those *belonging to the company* will be immediately followed by that of the officers and soldiers who since last
muster *have ceased to belong to it.* These will be classed in the following order, viz.: DISCHARGED, TRANSFERRED, DIED, DESERTED; and
the *utmost particularity* will be observed in the *remarks,* concerning
them; DATE and PLACE will *in every case* be given, and *No., date, &c.*
of *orders,* or *description of authority,* be always carefully specified.
Soldiers discharged *and re-enlisted,* or who have deserted *and been retaken,* since last muster, have their place in *both* of the above rolls.

Note 8.—The remark "*discharge and final statements given*" will be
made opposite to the name of every discharged soldier to whom such
papers *have actually been given.* But the blank spaces under the head
of LAST PAID, are to be filled as usual.

Note 9.—In all cases of "*re-enlistment*" prior to the expiration of the
term of service, the *discharge* on the old enlistment will be given at
the time the soldier "re-enlists"—from and on which day his pay on
the *new* enlistment will commence.

Note 10.—Within *three days* after each regular muster, the mustering
officer, or commandant of the post, will transmit to the Adjutant-General
a copy of the *Muster Roll* of each company. Blanks will be supplied
from the Adjutant-General's Office, and will be *acknowledged* on the
first Muster Roll forwarded after their receipt.

Note 11.—*Actions* in which the company, or any portion of it, has
been engaged, *scouts, marches, changes of station, every thing of interest,*
relating to the *discipline, efficiency, or service* of the company, will be
minutely and carefully noted, with DATE, PLACE, DISTANCES MARCHED,
&c. &c.

75. *Note* 1.—When the order is received promoting or
appointing an officer, or transferring him to the company,
he is taken up with the remark, " Promoted, or appointed, or
transferred to company per Special Order No. —, dated War
Department, Dec. 15, 1862, vice Lieut. A. B., promoted,
transferred, died, dismissed, or deserted," as the case may
be. At the foot of the roll the officer's name will be
entered under the proper heading, before the name of
similar cases of soldiers who have ceased to belong to the
company. Soldiers are taken up and dropped in a similar
way, always stating the number of order, whence issued, and
date.

76. *Note* 2.—There is a uniform way of making the remarks indicated in this note. The following will cover the majority of cases, viz. :—

Assigned to company, Jan. 1, 1863, per Special Order No. 1, dated War Department, Jan. 1, 1862.

Joined, Jan. 1, 1863, from leave of absence.

Assumed command of post, Jan. 16, 1862.

Relieved from command of post, Jan. 20, 1863, per General Order No 10, dated Head-Quarters Department of the West, Jan. 18, 1863.

On special duty as topographical engineer since Jan. 15, per Order No. 3, dated Fort Scott, Kansas, Jan. 6, 1863.

On extra duty as teamster in Quartermaster's Department since Jan. 15, 1863.

On daily duty as clerk in adjutant's office since Jan. 1, 1863, per Special Order No. 9, dated Head-Quarters, Fort Scott, Kansas, Jan. 1, 1863.

On daily duty as company cook since Jan. 15, 1863.

Promoted captain, per Special Order No. 10, dated War Department, Jan. 15, 1863.

Appointed corporal Jan. 1, 1863, per Regimental Order No. 15, dated Head-Quarters 2d U. S. Infantry, Fort Smith, Jan. 8, 1863.

Promoted sergeant, per Regimental Order No. 5, dated Head-Quarters 5th U. S. Infantry, Fort Smith, Arkansas, Jan. 10, 1863.

Reduced from sergeant, Jan. 7, 1863, per Regimental Order, dated Head-Quarters 4th U. S. Infantry, Fort Vancouver, W. T., Jan. 9, 1863.

Reduced from corporal, Jan. 10, 1863, per Order No. 5, dated Head-Quarters 5th U. S. Infantry, Jan. 15, 1863.

Due United States for ordnance, 25c.; for horse equipments, $25 00; for extra clothing, $2 90.

Paid for apprehension from desertion, $30.

Due paymaster, 25c., amount overpaid on Muster Roll for November and December, 1862.

Due laundress, $1 75.

Forfeited all pay due, by desertion, to Jan. 10, 1863, date of apprehension.

Due United States, $3, per sentence of regimental court-martial, per Order No. 5, dated Head-Quarters 2d U. S. Infantry, camp on Platte River, N. T., Jan. 7, 1863.

The stoppages are made in the order here given.

On sick leave since Jan. 3, 1863, for two months, per Special Order, dated War Department, Dec. 20, 1862.

Sick in general hospital, St. Louis, since Dec. 10, 1862.

On detached service since Jan. 10, 1862, per order No. 1, dated Head-Quarters Department Missouri, St. Louis, Dec. 10. 1862.

Wounded in battle, Antietam, Md., Sept. 14, 1862.

Sick in hospital.

Confined.

In confinement undergoing sentence.

77. *Note* 3.—The order in which stoppages shall be made is given above. No stoppages can be made against a soldier's pay unless authorized by law or regulations.

78. *Note* 4.—Section 2, Act of Aug. 4, 1854, gives to a soldier who re-enlists within one month after having been honorably discharged from the service, $2 per month, in addition to the ordinary pay of his grade, for five years after such re-enlistment, and $1 per month additional for each successive period of five years' service, so long as he remains continually in service. Section 3 gives $2 per month to those private soldiers who have received certificates of merit for services in the Mexican war, under act of March 3, 1847, section 17.

Section 4 gives to non-commissioned officers who were

recommended for promotion under the act of March 3, 1847, section 17, but did not receive the appointment, $2 per month in lieu of commission. The notes show the manner of making the remarks.

79. *Note* 5.—The bounty allowed by the act June 17, 1850, only applies to those soldiers who enlisted previous to July 1, 1861, in the regular army. Those who have enlisted since, if they serve two years or more, are entitled to $100. If any portion of the $100 has been advanced, it must be entered in the column headed BOUNTY PAID, and deducted in their final settlement. The regulations (Reg. 1857, Par. 1192 and 1193) are sufficiently clear and explicit on the bounty allowed previous to July 1, 1861, under the act June 17, 1850.

80. *Note* 6.—This note is sufficiently explicit.

81. *Note* 7.—To comply with this note, it is necessary to have the previous Muster Roll, and all those who have ceased to belong to the company within the period of muster must be accounted for as required by this note. A soldier who is reported on the last roll as DESERTED, if retaken or he has surrendered himself, is taken up in proper place alphabetically, with the remark "Joined from desertion," after which succeed the remarks according to the nature of the case.

82. *Notes* 8, 9.—Note 8 is sufficiently clear, except that the date must not be omitted. Note 9 relates to act of July 5, 1838, section 29, which allows the three months' extra pay alluded to in note 6, and allows the soldier to enlist within one month after, or two months before the expiration of his enlistment.

83. *Note* 10.—Gives three days to perfect the rolls; yet mustering officers who do not belong to the post or command, and have many troops to muster, could hardly allow so much time.

84. *Note* 11.—The record required in this note should not be omitted, as a matter of justice to the company. There is no specified form for it. It should be simply a plain statement of what the company has been doing in the two months, in which the dates, places, and distances marched are important.

85. Where the space is limited, the remarks given in Par. 76 may be abbreviated, for which purpose the list of abbreviations in this book should be used.

86. The *Recapitulation* should correspond with the face of the return. It represents in numbers what is specified in detail on the roll. The receipt of blanks should be regularly accounted for in the proper columns.

87. The roll which is retained, is produced when the paymaster arrives, and the pay and stoppages as computed by the paymaster are copied on it. The three rolls are then signed by the men before payment. When a name is not signed by the soldier himself, it must be witnessed by the officer who witnesses the payment. An officer of the company must always be present when the men are paid. They are usually paraded with their side-arms on : the officer calls the names in the order on the roll, and the first sergeant sees that the right men come forward to get their pay. The sutler is authorized to be present at the pay-table, and each soldier as he receives his pay steps around to the sutler's table and pays his account. The sutler's account should be completed and acknowledged beforehand, and only for the months for which the men are paid : this will prevent any dispute of the accounts at the pay-table.

88. In order to make out the Muster Roll correctly, it is necessary to have, first, the retained roll of the previous muster; the orders causing all important changes in the company in the two months; the amount of clothing made up of each soldier who has overdrawn his account; a record of all fines imposed by court-martials on the men of the company, and also of all charges to be made against the men for public property that they are to be charged with.

89. The remarks with regard to *discipline, instruction,* arms, &c., are made by the inspecting and mustering officer.

90. On the day of muster a check roll must be prepared of all the men who are present with the command but absent from parade, for the convenience of the mustering officer. It will include, ordinarily, those on guard, in confinement, sick in hospital, and those necessarily absent on extra duty; although these last should, if possible, be present to answer to their names.

Report of Damaged Arms.

91. This report is made after each muster, and is intended for the commanding officer of the regiment, corps, post, or garrison, to enable him to make the return required of him by the regulations of the Ordnance Department. (Reg. 1395.)

92. It is made out according to Form 6, and should state the nature of the damage to all arms, accoutrements, and implements, noting those occasioned by negligence or abuse, and naming the officer by whose negligence or abuse the damages were occasioned.

93. This return is intended to show the character of the damages to which arms, accoutrements, and implements are liable in the hands of troops, in order that the Ordnance Department may correct the defects in the manufacture. It is really an important report, although much neglected. Where a great variety are in use throughout the army, it gives authentic statistics, showing the defects of the various arms in use.

Par. 92.

Form 6.

Report of Damaged Arms of Company A, 1st U. S. Infantry, for the months of January and February, 1863.

Articles	Number		No. last report		Total.	Nature of damage.
	Unserviceable and irreparable.	Unserviceable but reparable.	Unserviceable and irreparable.	Unserviceable but reparable.		
Muskets, cal .58........	1	....	1	....	⎫ 7	⎰ Barrels bursted. ⎱ 2 stock broken, 2 mainspring broken, 1 hammer broken.
"	....	3	....	2	⎭	
Bayonets........	....	1	....	2	8	2 clasps broken, 1 broken at the shank.
Cartridge-boxes........	....	1	....	2	8	Stitching ripped.

Note.—The kind of arm in use should be stated, and where manufactured. The causes should appear, whether the result of accident or defect in the arm or accoutrement.

A—— B——, *Captain Co. A, 1st U. S. Infantry.*

Return of Clothing, Camp and Garrison Equipage.

94. The blanks for this return are furnished by the Quartermaster's Department the same as the Receipt Rolls. (See Form 51, Q. M. Reg., Gen. Reg.) They are made out quarterly in quadruplicate, one to be retained, one to be sent to department head-quarters (department head-quarters may not always require this), and the other two to the Quartermaster-General, one for an office copy, and the other for the third auditor, supported by the invoices of all the articles received, and the receipts for all issues and transfers.

95. Each supply of clothing, camp and garrison equipage, should be accompanied by triplicate invoices, for which corresponding receipts should be given. If the packages, when received, look as if they had been damaged, interfered with, or deficient in any respect, a board of survey should be applied for to the commanding officer. The board ascertains what is missing or damaged, and fixes the responsibility of the loss or damage. A copy is furnished to the officer, sending the stores and the officer receiving them, and a copy is retained at head-quarters of the commanding officer. The packages will, if they look perfect, be receipted for for what they are said to contain, without breaking them, and they will not usually be broken except for issue, and then if found deficient the board is called for as above. When the invoices are not numerous, they can be entered separately on the return, otherwise they should be consolidated; and the same with the receipts upon abstracts. The abstracts are similar to E and M, Forms 26 and 45, Q. M. Reg., Gen. Reg., by changing the headings.

96. Clothing cannot be dropped from the return in any other way than as issued or transferred, and must be supported by receipt rolls for the issues, and receipts from officers for the transfers. Damaged clothing may be issued at reduced rates, assessed by a board of survey, Reg. 1019. Clothing lost, deficient, or destroyed beyond the investigation of a board of survey, can only be accounted for by affidavit.

97. The law requires the officer " to show by one or more depositions, setting forth the circumstances of the case, that the deficiency was from unavoidable accident, or loss in active service, without any fault on his part, and, in case of damage, that due care and attention were exerted on his part, and that the damage did not result from neglect." These depositions accompany the return as vouchers.

98. Camp and garrison equipage is accounted for on the same return with the clothing, but in a different manner. It is permanent property for the use of the company, and is borne on hand as long as it is serviceable. When it becomes unserviceable, it is reported to the commanding officer, who applies for an inspector to examine it, and he makes an inspection report and recommends what shall be done with it. It is not dropped from the return until it has been inspected and ordered to be dropped. See Inventories and Inspection Reports, Par. 235.

99. Inventories are made of the property to be condemnéd, and handed in to the adjutant's office. These inventories state the articles, their number, from whom received, and how long they have been in service. They are then turned over to the inspector when he arrives, who makes his remark under the heading " Condition, &c.," and " Disposition." The inspector forwards the reports to the approving officer, who gives the orders in the case.

100. Articles which are issued to the soldiers for their constant use are charged to them on a list kept by the orderly sergeant, and, when lost or destroyed by their neglect, are entered against their pay on the Muster Roll, and deducted from it by the paymaster. (See register of articles issued, Par. 41.) Such are haversacks, canteens, talmas, bed-sacks, letters, numbers, cross-sabres, bugles, &c. &c. If the company—as it should be—is divided into squads under the sergeants, the tents, axes, spades, shovels, cooking-utensils, &c. are proportionately distributed to them, and they will be held responsible for them. If the articles are lost or injured by the neglect of any one, they (the sergeants) must report to whom they are to be charged, or else they will be charged to themselves.

101. It is advisable not to keep on hand any more property than is necessary for the immediate wants of the company.

102. The following are forms of invoices and receipts of clothing, camp, and garrison equipage.

Invoice of clothing, camp and garrison equipage, this day turned over to Captain —— ——, 2d Ohio Vol. Cavalry, by Lieutenant —— ——, First Lieutenant and Regimental Quartermaster, viz.:

NUMBER.	ARTICLES.
40	Forty Great-Coats.
50	Fifty Trowsers.
10	Ten Forage Caps.
5	Five Blankets.
7	Seven Tents.

H—— A——,
1st Lt. and Reg. Q. M.
2d Ohio Vol. Cav.

FORT SCOTT, KANSAS, Nov. 10, 1862.

Received this day, of First Lieutenant H—— A——,
Regimental Quartermaster 2d Ohio Vol. Cavalry, the fol-
lowing articles of clothing, camp and garrison equipage,
&c., viz. :—

NUMBER.	ARTICLES.
40	Forty Great-Coats.
50	Fifty Trowsers.
10	Ten Forage Caps.
5	Five Blankets.
7	Seven Tents.

A—— L——.

Capt. 2d Ohio Vol. Cav.

FORT SCOTT, KANSAS, Nov. 10, 1862.

103. These forms may be drawn up also the same as
Form 27, Q. M. Reg., Gen. Reg., with the necessary altera-
tion in the headings.

104. Where no invoices have been sent or received with
the property, certified invoices must be made out, stating
from what source they have been received. The certificate
must state that it is a correct list of the property received.
and that no invoices were received with it.

Clothing, Receipt Rolls.

105. These are made out in duplicate on printed blanks
obtained by requisition on the Quartermaster-General, or the
chief quartermaster of the department. If no printed
blanks are on hand, they must be made in manuscript according to Form 52, Q. M. Reg., Gen. Reg. The number
of articles drawn under each heading are entered opposite
the soldier's name, and the blank spaces filled with parallel

lines. These rolls are then signed by the soldier, and his signature, as well as the issues, are witnessed by an officer: if no officer be present, then a non-commissioned officer must witness them.

106. These receipt-rolls are the officer's vouchers, and one copy is retained by him, and the other accompanies his return of clothing, camp and garrison equipage. These receipt-rolls should be completed at the time of issue, and signed and witnessed, as the soldier may be ordered away the next moment, and the issuing officer never see him again.

Return of Ordnance.

107. Printed blanks are furnished on application to the Ordnance Department. Requisitions should be made by commanding officers of posts and regiments for their commands, stating the number of companies and the arm. When no blanks are furnished, they are usually, for a company, ruled out on a sheet of ordinary foolscap, which is generally large enough for the amount and kind of ordnance in the hands of a company commander. It is made out quarterly, viz.: at the end of March, June, September, and December, in triplicate, and should the department commander require it, a copy is made for his head-quarters. Two copies are sent to the chief of ordnance: one is intended for his office, and the other, with the vouchers, is sent by him to the second auditor, who audits the return. One copy of the return and of each of the vouchers is retained by the officer making it, in case the originals are lost, and for future reference.

108. A letter of transmittal should accompany the returns according to the following form:—

....

.... 186.

SIR :

I have the honor to transmit herewith the original and duplicate copies of the Return of the Ordnance and Ordnance Stores appertaining to ....

* ....

for which I am accountable for the .... quarter of 186 .

    Respectfully,

      Your obedient servant,

     ....

     ....

       Commanding ....

To the Chief of Ordnance,
   Washington, D.C.

  { If transmitting a Company or Regimental Return, give letter of
 * { Company and designation of Regiment.
  { If a Fort or Battery Return, give the official designation of the Post.

109. On the lines "received from," are entered the names of the officers from whom the property was received, and the number of articles on the invoices are entered opposite on the same line, under the respective headings. Should the invoices be numerous, they may be consolidated on an abstract, Form 7.

110. On the lines "issued to" are entered the names of the officers to whom ordnance has been transferred; also the abstract of losses, expenditures, and charges. If the issues, &c. are numerous, they are also consolidated on abstracts similar to Form 7.

PAR. 109.

Form 7.

Abstract of Ordnance received by Captain A—— B——, 1st U. S Infantry, in the quarter ending Dec. 31, 1862.

Date.	Number of invoice.	From whom received.	Muskets, cal. .58.	Bayonet-scabbards.	Cap-pouches and picks.	Cartridge-boxes, cal. .58.	Cartridge-box plates.	Cartridge-box belts.	Cartridge-box belt-plates.	Gun-slings.	Waist-belts, private.	Waist-belt plates.	Screw-drivers and wrenches.	Wipers.	Tompions.	Blank-cartridges.	Ball-cartridges.	Packing boxes.
		Total received...........																

A—— B——,
Captain 1st U. S. Infantry.
Commanding Company.

Par. 111.

Form 8.

(For the use of Cavalry and Infantry.)

Abstract of Materials, &c., expended or consumed in Company ———, ——— Regiment ——— during the ——— quarter, 186 .

DATE.	HOW EXPENDED.	CLASS VIII.—AMMUNITION.							CLASS IX.	MATERIALS.
		Elongated Ball Cartridges, calibre.........	Ball Cartridges, calibre.........	Carbine Cartridges, calibre.........	Pistol Cartridges, calibre.........	Blank Cartridges.	Percussion Caps.			
186 .	In practice firing.............									
	In action at.................									
	In action at.................									
	In action at.................									
	In the repair of arms.........									
	In repair of accoutrements and equipments....									
	TOTAL EXPENDED.............									

I certify that the above abstract is correct, and that the stores have been expended for the purposes stated.

[IN TRIPLICATE.]

....
Commanding.

NOTES. { No other stores than ammunition and materials can be expended on this abstract.
Give letter of company, the Regiment, State, and arm of service.
If more headings are required, gum a piece of paper on the right-hand edge, ruled in conformity with this sheet.

111. ABSTRACT OF ORDNANCE EXPENDED (see forms in instructions) Form 8.—On this is entered the ammunition, fuses, caps, port-fires, &c. used in service. Arms, horse-equipments, &c. cannot be entered on this abstract as expended. When arms, accoutrements, &c. become worn out and unserviceable, they must be inspected and ordered to be dropped, and the Inventory and Inspection Reports become the vouchers for dropping them from the Return. (See par. 235.) Ordnance is seldom ordered to be dropped, but most generally to be turned into the Ord. Department.

112. When ordnance is lost in battle, or from other unavoidable cause, the following certified statement is the manner of accounting for the loss. (See instructions for making ordnance returns.)

CAMP NEAR MURFREESBORO', TENN.,

December 31, 1862.

I certify, on honor, that on the 30th day of December, 1862, at Murfreesboro', Tenn., the Stores enumerated below were lost under the following circumstances:—

The Regiment to which my company belongs was directed to advance under the fire of the enemy to take a certain position; in so doing, ten privates and two non-commissioned officers were killed, and ten privates and one non-commissioned officer were severely wounded.

·The arms carried by all these men were left on the field, as we were repulsed, and they could not be recovered. The following is the list of Stores so abandoned:—

23 Springfield rifled muskets, calibre .58.

12 Infantry cartridge boxes and plates.

12 Infantry cartridge-box belts and plates.

12 Infantry waist belts and plates.

12 cap pouches and picks.

12 gun slings.

10 ball screws.

1 spring vise.

JAMES G. BROWN,

IN DUPLICATE.   *Captain 250th Del. Vols.,*

———   *Commanding Company C.*

Endorsement to be as follows:—

No......

List of Stores lost in action

at

Murfreesboro', Tenn.,

December 30, 1862.

When any enlisted man or citizen is cognizant of the facts connected with any loss of the above nature, the officer should obtain
from him an affidavit to that effect as follows:—

The undersigned, being duly sworn, deposes and says that he is
cognizant of the above facts as above set forth, and that they are
correct to the best of his knowledge and belief.

HENRY F. WILSON,

Private, Co. C, 250th Del. Vols.

*Sworn to and subscribed before me, at Murfreesboro', Tennes-
see, this 31st day of December, 1862.

P. C. COWEN, J. P. [L. s.]

113. ABSTRACT OF CHARGES.—On this abstract are
entered the charges made against the enlisted men for the
loss of arms, accoutrements, &c. Opposite the names,
under the respective headings, are entered the articles, and
then the muster-roll on which the articles were paid for;
or, if not paid for, the muster-rolls on which they are
charged, Form 9. (See statement of the prices of horse-
equipments and small arms, appended, General Order No.
202, Dec. 9, 1862.) Reg. 1393. (See par. 57 to 59 instructions for making ordnance returns, also, page 114 for
a form which may be used instead of Form 9.)

114. The invoices are made out agreeably to Form 2,
Gen. Reg. Ord. Dep't. In turning over ordnance from one
company to another which is in the hands of the soldier,
this form requires some little modification. Under any
circumstances Form 10 is preferable, as it will answer
whether the ordnance is packed or not.

---

* See paragraph 1031, Revised Regulations of the Army.

Par. 113.

Form 9.

Abstract of Ordnance charged to Soldiers of Company A, 1st U. S. Infantry, in the Quarter ending March 31, 1863.

To Whom Charged.	Muskets.	Bayonet-scabbards.	Cap-pouches and picks.	Cartridge-boxes.	Cartridge-box plates.	Cartridge-box belts.	Cartridge-box belt-plates.	Gun-slings.	Waist-belts.	Waist-belt plates.	Screw-drivers.	Wipers.	Tompions.	On What Muster Rolls.
Private John Brown......	1	1	1	1	1	1	1	1	1	1				Jan. and Feb 1863, deserted.
" John Smith......											1	1	1	March and April.
" James Johnson......			1					1			1	1	1	January and February.

I certify that the above Abstract is correct; that the articles were charged on the Muster Rolls as stated.

A—— B——, *Capt. Co. A, 1st U. S. Infantry.*

PAR. 114.

Form 10.

Invoice of Ordnance and Ordnance Stores this day turned over to Capt. A—— B——, 1st U. S. Infantry, by Lieut. J—— C——, 1st U. S. Infantry, viz. :

Number.	Articles.	Condition.	Weight.	Marks on Packages.	No. of Boxes or Packages.
80	Eighty muskets complete, cal..58	Serviceable.			
6	Six non.-com. swords and belts	Worn.			
80	Eighty bayonet-scabbards........	Worn.			
80	Eighty cap-boxes and picks	Worn.			
80	Eighty cartridge-boxes	Worn.			
4	Four " "	Unserviceable.			

J—— C——, 1st Lieut. 1st U. S. Infantry.

Statement of the Cost of Horse Equipments, pattern 1859.

NAMES OF PARTS.	Price per Piece.	Price per set.	Amount.
SADDLE.			
Saddle tree, covered with raw hide, with metal mountings attached	$3 87	$3 87	
Saddle flaps, with brass screws, each	1 18	2 36	
Back straps, with screws, rivets, and D's, each	52	1 04	
Girth straps, long	36	36	
" " short	23	23	
Cloak straps, each	17	1 02	
Stirrup leathers, each	57	1 14	
Sweat leathers, each	30	60	
Stirrups, with hoods. each	38	76	
Carbine socket and strap	47	47	
Saddle bags	3 50	3 50	
Crupper	1 01	1 01	
Girth	66	66	
Surcingle	1 16	1 16	
Total cost	....		$18 18
BRIDLE.			
*Bit, No. 1, $3 50 .... } average per 100 sets	2 94	2 94	
†Bit, Nos. 2, 3, and 4. $2 80... }			
Brass scutcheon, with company letter, each	5	10	
Reins	55	55	
Head piece	67	67	
Front	16	16	
Curb chain, with hooks	14	14	
Curb chain safe	4	4	
Total cost	....		4 60
HALTER.			
Head stall, complete	1 55	1 55	
Hitching strap	48	48	
Total cost	....		2 03
WATERING BRIDLE.			
Snaffle bit, chains, and toggles	50	50	
Watering rein	56	56	
Total cost	....		1 06
Spurs	20	40	
Spur straps	10	20	
Total cost	....		60
Currycomb	20	20	
Horse brush	67	67	
Picket pin	13	13	
Lariat rope	61	61	
Total cost	....		1 61
Total cost of equipment	....	....	28 08
Blanket for cavalry service, dark, with orange border, 3 lbs., at 75 cents per lb	2 25	2 25	
Blanket for artillery, scarlet, with dark blue border, 3 lbs., at 70 cents per lb	2 10	2 10	
Nose bag	1 25	1 25	
Hitching strap	25	25	

* NOTE.—No. 1 is Spanish, Nos. 2, 3, and 4 are American.
† NOTE.—For officers' scutcheons, gilt, $0 15 each.

*Table showing the Prices of Malleable Iron Parts, Buckles, D's,
Rings, &c.*

Tabular No. of piece.	Place where used, and kind of buckle.	Number required in each set.	Size.	Price.
			Inch.	Cts.
1	Girth, with roller, round.............	1	2	2
2	Stirrup, bar, flattened................	2	1.375	2
3	Halter, bar, flattened................	1	1.125	2
4	Girth and surcingle, roller, round	2	1.5	2
5	Bridle, crupper, bar....................	4	.75	1
6	Throat lash, saddle bags, cloak straps, and carbine socket, bar.	12	.625	1
7	Halter, square..........................	2	1.6✕1.2	2
8	Halter ring..............................	2	1.7	2
9	Ring for crupper and saddle tree..	5	1.25	1
10	Halter bolt.............................	1	1.10	1
11	Foot staples............................	6	.9	1
12	D's, back straps, and girths........	3	1.85	4
13	Saddle bags' stud.....................	1	1✕0.4	2

Prices of Small Arms.

Names of Parts.	U.S. Muskets, smooth bore, Calibre 69	U.S. Muskets, rifled, Calibre 69	U.S. Rifle Musket, Springfield model, Calibre 58	U.S. Rifles, Harper's Ferry model, Cal. 54 and 58	Foreign smooth bore Muskets, all calibres	Foreign Rifle Muskets, Enfield model	Other Foreign Rifled Muskets, all calibres
Barrel	$4 10	$4 67	$5 70	$5 26	$2 19	$5 52	$3 80
Breech screw	10	10	41	10	04	10	07
Vent screw			03	03	01	03	02
Cone	09	09	12	09	03	10	07
Front sight	01	01	02	02	01	02	02
Rear sight — Base		38	38	38	15	40	28
Rear sight — Base screw		05	05	05	02	05	03
Rear sight — First leaf		22	22	22	09	23	16
Rear sight — Second leaf		25	25	25	10	26	18
Rear sight — Joint screw		03	03	03	01	03	02
Barrel and attachments, complete	4 30	5 80	7 21	6 43	2 65	6 74	4 65
Stock, complete	$1 45	$1 45	$2 00	$1 85	$0 90	$1 88	$1 45
Upper band	$0 28	$0 28	$0 15	$0 15	$0 06	$0 12	$0 10
Middle band	18	18	30				
Lower band	15	15	16	18	07	14	12
Band swivel	12	12	12	12	05	09	08
Band springs, 10 cts. each	30	30	30	30	13	23	21
Guard plate	42	42	50	50	21	39	34
Guard bow	30	30	38	35	14	26	24
Guard-bow nuts, 2 cts. each	04	04	04	04	02	03	02
Guard-bow swivel	10	10	10	10	04	08	07
Trigger	12	12	20	12	05	09	08
Trigger screw	02	02	03	02	01	02	02
Guard screws, 4 cts. each	08	08	08	08	03	06	05
Butt plate	30	30	46	53	22	41	36
Butt-plate screws, 4 cts. each	08	08	08	08	03	06	05
Side plate	07	07		10	04	08	07
Side-screw washers, 4 cts. each			08				
Tip of stock			18	18	07	14	12
Tip screw			03	03	01	02	02
Tang screw	05	05	05	05	02	04	03
Side screws, 7 cents each	14	14	14	14	06	11	09
				74	31	58	50
Grease box — Plate				05	02	04	03
Grease box — Catch				10	04	08	07
Grease box — Spring				02	01	02	02
Grease box — Spring screw				09	04	07	06
Grease box — 3 box screws							
Mountings, complete	2 75	2 75	3 88	4 07	1 68	3 16	2 75

Prices of Small Arms.—Continued.

	NAMES OF PARTS.	U.S. Muskets, smooth bore, Calibre .69.	U.S. Muskets, rifled, Calibre .69.	U.S. Rifle Musket, Springfield model, Calibre .58.	U.S. Rifles, Harper's Ferry model, Calibres .54 and .58.	Foreign smooth bore Muskets, all calibres.	Foreign Rifle Muskets, Enfield model.	Other Foreign Rifled Muskets, all calibres.
Lock.	Plate	$0 50	$0 50	$0 80	$0 50	$0 31	$0 71	$0 50
	Hammer	60	60	66	60	37	86	60
	Tumbler	27	27	50	27	17	39	27
	Tumbler screw	03	03	03	03	02	04	03
	Bridle	16	16	22	16	09	23	16
	Bridle screw	03	03	03	03	02	04	03
	Sear	20	20	51	20	12	29	20
	Sear screw	03	03	03	03	02	04	03
	Sear spring	10	10	16	16	06	14	10
	Sear-spring screw	03	03	03	03	02	04	03
	Main spring	27	27	50	27	17	39	27
	Main-spring swivel			12				
Lock, complete		2 22	2 22	3 39	2 22	1 37	3 17	2 22
Ramrod		$0 50	$0 50	$0 75	$0 50	$0 32	$0 57	$0 50
Ramrod spring		12	12		12	07	14	12
Ramrod stop		01	01	02	01	01	01	01
Ramrod, complete		63	63	77	63	40	72	63
Bayonet		$1 47	$1 47	$2 25		$0 90	$2 10	$1 17
Bayonet clasp		16	16	22		09	22	11
Bayonet-clasp screw		02	02	03		01	01	02
Bayonet, complete		1 65	1 65	2 50		1 00	2 33	1 30
Sword Bayonet.	Blade				$1 40			$1 40
	Hilt				1 75			1 75
	Lock pin				08			08
	Lock-pin spring				05			05
	Lock-pin screw				02			02
Sword bayonet, complete					3 30			3 30
Arm, comp'te.	With bayonets	$13 00	$14 50	$19 25		$8 00	$18 00	$13 00
	With sw'd bayonets				$18 50			15 00
	Without bayonets				15 20			

Appendages for all Arms.

Spring vice, 30 cents each; 1 to 10 arms... 03 cents.
Ball screw... 13 "
Wiper.. 20 "
Screw-driver .... 23 "
Tompion.. 02 "
Tumbler and band spring punch, 20 cents each; 1 to 10 arms..... 02 "
Spare cone.. 12 "
                                                      75 "

Prices of Sharp's Rifle and Carbine.

NAMES OF PARTS.	RIFLE. Price.	CARBINE. Price.
Barrel	$6 00	$4 00
" Spline	75	....
" Stud..	25	25
" Bouching	50	50
Front sight-stud...	08	08
" " pin..	02	02
" " " silver...	05	05
Rear sight spring..	33	33
" " " screw..	05	05
" " pin..	02	02
" " leaf...	40	40
" " base...	25	25
" " slide..	15	15
" " " screw..	05	05
Receiver	3 00	3 00
Slide...	1 75	1 75
" screw...	05	05
Gas plate...	75	75
Two cones, 10 cents each...	20	20
Slide vent tube..	10	10
" " " screw...	05	05
Toggle..	25	25
Lever...	1 00	1 00
" screw...	05	05
" spring..	15	15
" " screw	05	05
" " key...	05	50
" key stop...	05	05
" " " spring...	06	06
" " " " screw..	05	05
" " " " pin..	05	....
" catch...	40	40
" " screw	05	05
" " spring..	10	10
" " " screw..	05	05
" " " pin..	05	05
Guard plate...	1 00	1 00
Two guard-plate screws, 5 cents each...........................	10	10
Trigger...	30	30
" screw...	05	05
Lock plate	1 75	1 75
Hammer...	1 00	1 00
Tumbler..	60	60
" screw..	05	05
" stirrup..	08	08
" " screw..	05	05
Main spring...	56	56
" " screw	05	05
Sear	30	30
" screw...	05	05
Bridle..	30	30
Two bridle screws, 5 cents each.................................	10	10
Amount carried forward...	$28 55	$21 29

Prices of Sharp's Rifle and Carbine.—Continued.

NAMES OF PARTS.	RIFLE. Price.	CARBINE. Price.
Amount brought forward..	$23 55	$21 20
Primer spring	08	08
" " screw..	05	05
" follower..	05	05
" " pin..	02	02
" driver...	25	25
" slide..	15	15
" cover...	10	10
" " screw..	05	05
" " pin..	02	02
" shut-off	25	25
" " screw......................................	05	05
Two side screws, 5 cents each..............................	10	10
Two tang screws, 5 cents each..............................	10	10
Stock butt..	2 25	2 25
Butt plate..	1 00	1 00
Two butt-plate screws, 5 cents each.......................	10	10
Patch-box head...	37	37
" lid	37	37
" joint pin..	04	04
" spring..	10	10
" screw..	05	05
Three patch-box head-screws, 5 cents each..............	15	15
Patch-box hook...	12	....
Two patch-box pins, 2 cents each..........................	04	....
Stock tip...	1 75	75
" screw..	05	05
" escutcheon...	05	05
Nose cap	25	....
" screw..	05	....
Upper band..	45	45
Middle band and swivel screw..............................	1 25	....
Lower band	50	....
Band springs, 10 cents each	30	10
Swivel on stock butt..	75	....
Swivel bar...	....	40
" ring..	....	10
" screw..	....	05
Bullet mould, $1 50 each; 1 to five arms.................	....	30
Bullet mould	30	....
Cone wrench and screw-driver..............................	37	37
Brush thong and rod..	....	48
Brush	25	....
Thong	15	....
Rod	15	....
Cartridge stick..	05	....
Cost of arm...	36 08	30 00
Sabre and scabbard	5 00	....
Guard plate, with double trigger, complete...............	2 75	....
Angular bayonet..	2 50	....

Prices of Merrill's Rifle and Carbine.

NAMES OF PARTS.	RIFLE. Price.	CARBINE. Price.
Barrel	$8 00	$6 00
Breech	5 50	4 83
Lever...	3 50	3 50
" catch	50	50
" " spring and screws.....................................	10	10
Link	90	90
Plunger..	1 50	1 50
Two link screws, 5 cents each............................	10	10
Rear sight..	1 00	1 00
Stock..	5 50	3 50
Guard, trigger, and screws.................................	1 25	25
Side plates	60	60
Patch box	75	....
" spring and screw...............................	13	....
Butt plate..	80	80
Bands, 25 cents each..	50	25
Two nipples and wrench......................................	50	50
Lock plate..	37	37
Hammer	50	50
Main spring..	25	25
Sear spring...	12	12
Tumbler	38	38
Sear..	25	25
Bridle	25	25
Lock screw...	13	13
Two swivels, 40 cents each..................................	80	80
Nose cap..	25	25
Two barrel catchers and screws, 37½ cents each.............	75	75
Steel wash rod...	50	50
Brass wiper..	12	12
Sabre bayonet and scabbard................................	5 20	....
Cost of arm complete....................	$41 00	$30 00

Prices of Carbines.

NAMES OF PARTS.	Smith's.	Burnside's.	Gallager's.
Barrel..	$6 00	$5 00	$9 14
Frame...	....	....	9 00
Barrel receiver...	1 95	....	....
Chamber......	....	3 00	....
" bouching..	....	10	....
" screws, 10 cents each............................	....	20	....
Breech piece...	4 00	4 10	....
" pin...	....	1 75	....
" bolt..	20	05	....
Amount carried forward........................	$12 15	$14 20	$18 14

Prices of Carbines.—Continued.

NAMES OF PARTS.	Smith's.	Burn-side's.	Galla-ger's.
Amount brought forward........................	$12 15	$14 20	$18 14
Breech bolt spring...............................	....	03	....
Lever	....	....	1 25
Lever catch	....	....	10
Latch..	....	75	....
" spring....................................	....	10	....
" screw....................................	....	10	....
Clasp spring......................................	84	....	....
" " lifter..............................	22	....	....
Link...	....	....	60
Link slide..	....	....	60
Trap cover...	....	10	....
Cover spring	....	06	....
Hook..	....	12	....
Front sight..	75	10	06
Rear sight..	1 44	62	50
Barrel stud...	....	10	....
Stock bank...	75	....	....
Lock plate...	....	50	63
Hammer..	90	37	50
Tumbler..	75	30	30
Main spring..	56	37	25
Bridle...	....	25	20
Stirrup..	22	10	10
Trigger...	45	20	20
" spring......................................	....	08	....
Sear...	....	25	25
" spring.......................................	75	....	....
" " pin....................................	06	....	....
Cone..	14	10	....
Stock...	....	....	2 50
" butt..	1 80	2 50	....
Barrel stock.......................................	60	50	....
Band..	25	25	....
" spring..	....	10	....
" swivel..	50	50	....
Ring bar..	30	20	15
Ring...	15	10	05
Butt plate..	75	50	20
Side plate..	75	....	....
Guard plate..	....	35	....
" bow..	1 00	2 50	....
" plate stud...................................	10	....	....
" joint screw..................................	....	10	....
Patch box..	....	....	40
Top plate...	....	25	40
Bottom plate.......................................	....	....	1 00
Escutcheon	10	10	....
Small screws. 5 cents each	75	1 25	1 25
Bullet mould.......................................	1 25	1 25	....
Cone wrench.......................................	40	40	37
Brush thong and wiper	35	33	....
Tompion	....	02	....
Cost of carbine...................................	$29 00	$30 00	$30 00

Carbines, complete.

Maynard's, $30.   Joslyn's, $30.   Lindner's, $30.

Prices of Revolver Pistols.

NAMES OF PARTS.	Colt's Army.	Colt's Belt.	Remington's.
Barrel	$7 00	$6 10	$2 55
Barrel stud			10
Sight	03	03	07
Stock frame	4 50	3 90	4 66
Base pin	31	31	
Centre pin			47
Centre-pin spring			05
Bolt	29	29	33
Cylinder	4 46	3 86	2 75
Seven cones	39	39	56
Lever	1 33	1 33	93
Lever catch	05	05	08
Lever-catch spring			04
Hook	09	09	
Link			08
Link screw	04		
Key	28	28	
Key spring	09	09	
Hand	28	28	
Hand spring	02	02	
Pawl			20
Cam			07
Roll	03	03	07
Rammer	35	35	13
Hammer	71	71	60
Main spring	33	33	27
Sear spring	08	08	07
Stock	53	46	44
Stock strap	1 11	97	
Guard	28	28	
Trigger	1 24	1 09	27
Small screws, 2 cents each	28	28	16
Rivets, one cent each	04	04	
Cost of pistols	$24 14	$21 64	$14 95
Appendages.			
One six-bullet mould to 50 pistols, $4 50 each	09	09	
One two-bullet mould to 2 pistols, 80 cts. each	40	40	
One single-bullet mould to each pistol			80
Screw-driver and cone wrench	37	37	25
Cost of pistol and appendages	$25 00	$22 50	$16 00

Revolver pistols, complete.

Starr's......... $25 00   Joslyn's......... $22 50   Savage's......... $20 00
Whitney's.... 15 00   Lefaucheux.... 14 00

Prices of Sabres and Swords.

NAMES OF PARTS.	SABRES.			SWORDS.		
	Cavalry, American.	Cavalry, Foreign.	Horse Artillery.	Non-com. Officers.	Musicians.	Foot Artillery.
Blade................................	$3 00	$2 25	$2 25	$2 20	$1 92	$2 13
Hilt. { Pommel	75	54	54	50	44	....
{ Gripe........................	22	15	15	24	20	87
Guard................................	1 20	87	87	1 20	44	....
Scabbard. { Body	1 28	93	93	66	50	50
{ Mouth-piece	20	14	11	....	....	....
{ Bands and rings.............	60	44	44	....	....	....
{ Ferrule and stud...........	....	....	....	35	25	25
{ Tip.............................	25	18	18	35	25	25
Cost of arm complete................	$7 50	$5 50	$5 50	$5 50	$4 00	$4 00

NOTE.—**Extra** parts are accounted for in the same way as the complete **arm**, and must be entered on the Return as invoiced. Any parts lost by soldiers are charged to them under the same circumstances as when they lose the complete arm.

Prices of Accoutrements.

PARTS.	Infantry.	Artillery.	Cavalry.	Rifle.
Cartridge box...	$1 10			$0 95
Cartridge box plate..	10		$0 10	10
Cartridge box belt..	69			
Cartridge box belt plate.................................	10			
Bayonet scabbard and frog............................	56			
Waist belt. private's.......................................	25			37
Waist belt plate...	10			10
Cap pouch and pick.......................................	40		40	40
Gun sling...	16			16
Sabre belt..		$1 03	1 35	
Sabre belt plate..		60	60	
Sword belt...		1 00		
Sword belt plate..		10		
Sword belt, non-commiss'd officer's and musician's..	62			62
Sword belt plate...........do....................do..........	10			10
Waist belt.................do....................do..........	37			37
Waist belt plate..........do....................do..........	60			60
Carbine cartridge box...................................			87	
Pistol..............do...			75	
Holsters, with soft leather caps....................			2 63	
Carbine sling..			95	
Carbine swivel..			88	
Sabre knot..			30	
Bullet pouch...				53
Flask and pouch belt.....................................				40
Powder flask...				1 20
Waist belt. sapper's, with frog for sword bayonet, $1..				

Supplement to Prices of Small Arms.

United States rifle musket, calibre .58 ....$20 00
United States rifle, with sword bayonet, calibre .58................................ 18 50
United States musket, smooth bore, calibre .69.................................... 12 00
United States musket, rifled, calibre .69... 13 50
Colt's revolving rifle and bayonet.. 45 00
Merrill's breech-loading rifle and bayonet.. 45 00
Sharp's............do................do.. 42 50

FOREIGN MANUFACTURE—
    Austrian smooth bore... 6 00
    Prussian...........do... 6 00
    Belgian............do... 6 00
    English Tower...do... 6 00
    Austrian, rifled.. 10 00
    Prussian, rifled... 10 00
    Belgian, rifled... 10 00
    English Enfield rifles... 19 00
    French rifles... 16 00

CARBINES.—Sharp's, Gallagher's, Merrill's, Maynard's, Joslyn's, Smith's, and
    Lindner's... 30 00

REVOLVING PISTOLS.—Colt's, holster... 25 00
    Colt's, belt.. 22 50
    Savage's.. 20 00
    Starr's.. 20 00
    Remington's.. 16 00
    Whitney's... 15 00
    Le Faucheux's... 13 00

Cavalry sabres (American)... 7 50
Cavalry sabres (foreign)... 5 00

PAR. 117.

Form 11.

Quarterly Return of Quartermaster's Stores received and issued in Company "A," 1st U. S. Cavalry, in the Quarter ending March 31, 1863, by Capt. A—— B——, 1st U. S. Cavalry.

Date, 1863.	No. of invoice and voucher.	From whom received.	Horses.	Mules.	Wagons.	Wheel-harness, sets.	Lead-harness, sets.	Wagon-covers.	Wagon-saddles.	Saddle-blankets.	Halters and straps.	Wagon-whips.	Water-buckets.	Tar-buckets.	Blacksmith tools.	Portable forge.	Lanterns.	Currycombs.	Horse-brushes.	Picket-rope.	Lumber, feet.	Nails, pounds.	Remarks.
Jan. 15.	1	On hand last Return.....	50	6	1	1	2	1	1	2	6	1	1	1	1	1	4	2	2	1	1,000	100	
	1	Capt. A—— B——, Q. M.	10					1		1		1		1			1	1	1		500		
		Total to be accounted for	60	6	1	1	2	2	1	3	6	2	1	2	1	1	5	3	3	1	1,500	100	
Feb. 10.	1	To whom issued. Capt. A—— B——, Q. M.	9																				
March 30.	2	Dropped per inventory and inspection report...						1		1		1		1			1	1	1				Condemned.
March 31.	3	Expended per abstract.....																			1,200	50	
"	4	Lost per certificate.....	1																				
		Total issued.....	10					1		1		1		1			1	1	1		1,200	50	
		On hand to be accounted for.....	50	6	1	1	2	1	1	2	6	1	1	1	1	1	4	2	2	1	300	50	

I certify that the above Return exhibits a true and correct statement of all the property which has come into my hands on account of the Quartermaster's Department, during the quarter ending on the 31st of March, 1863.

A—— B——, Capt. Co. "A," 1st U. S. Cavalry.

115. When damaged arms begin to accumulate, an invoice of them is made and the nature of the repairs required stated. They are then packed and turned over to the Quartermaster's Department for transportation to the nearest arsenal. This is done under the direction of the commanding officer. When the arms are repaired, they are returned to the company. The packages are marked with the letter and regiment of the company, in addition to the address. The invoice accompanies the packages to the ordnance officer.

116. The Ordnance Department is very particular on the subject of endorsements, and the following circular is inserted for the instruction of officers making returns to this department, viz. :—

(CIRCULAR.)
ORDNANCE OFFICE,
WASHINGTON, May 6, 1862.

Hereafter, all letters or papers to be transmitted to this Office must be folded *three and a half inches wide,* agreeably to the requirement of Article 150, Ordnance Regulations, edition of 1852, except where otherwise specially directed, or in cases of voluminous papers not susceptible of being so folded.

The endorsement, required by the same article, will be omitted from the back of *letters.*

The attention of all officers corresponding with this office is particularly called to this subject.

JAMES W. RIPLEY,
Brigadier-General.

For more complete information on the subject of Ordnance Returns see Instructions for making Ordnance Returns, issued by the Ordnance Department.

Return of Quartermaster's Property.

117. Officers of artillery and cavalry commanding companies are generally responsible for a small amount of quartermaster's property, and sometimes infantry officers also, which must be accounted for on a separate return, and they must be careful not to include it with clothing, camp and garrison equipage, or ordnance. It must be made out on the general principles of Form 23, Reg. Q. M. Dep't, the property being generally of the same character and amount. Form 11 is given for their guidance.

118. The same general principles must be observed as have been explained for the returns of clothing, camp and garrison equipage, and ordnance, viz. :—

Invoices must always accompany the property received. They may be consolidated on an abstract ("E"), Form 26, Reg. Q. M. Dep't, or, if not numerous, may be entered on the Return separately. If no invoices are to be had, then certified invoices, signed by the officer who has received the property, must be made.

119. Receipts must be taken and invoices given to all property transferred, which may be entered separately, or, if numerous, consolidated on an abstract. (M, Form 45, Reg. Q. M. Dep't.)

120. Where the amount of property expended, or lost and destroyed, is small, and the number of certificates not numerous, they may be inserted on the property return, as in Form 11. Where the amount of property is great, Abstract "L," Form 41, and Vouchers 42 and 43, Q. M. Reg., must be complied with.

121. With regard to property worn out in public service, the inventory and inspection reports must be resorted to before the property can be dropped from the return.

122. The Quartermaster's Department supplies quartermasters with printed blanks for quartermaster's property; but it has not been the custom to supply them to company commanders, and heretofore they have always been made out in manuscript.

Quarterly Return of Deceased Soldiers.

123. This return is made out at the end of each quarter, in duplicate; one copy is retained, and the other is sent to regimental head-quarters, to enable the commanding officer of the regiment to make out a return for the regiment. Blanks for the purpose are furnished from the Adjutant-General's office.

124. The deceased soldier's name is entered in the first column, and opposite the name his rank and company, when and where he died, and the cause of his death. Under the heading "Due the Soldier" are entered the various amounts of pay due. First, his pay proper is expressed in months and days from the date of last payment.

125. Next, the retained pay, remembering that previous to the act of Aug. 3, 1861, the amount of pay retained was $1 per month, and subsequently $2; and by the act of July 17, 1862, it was again made $1, which law is still in force.

126. The extra pay due is entered as follows: First, what may be due him as hospital steward, remembering that at posts of four companies and more it is $22, and at other posts $20, per month. It is only the difference between his pay proper and what is allowed as hospital steward, that is entered in this column (Act July 5, 1838, sec. 12); Second, what is due him as nurse or cook in hospital (Act Aug. 16, 1856, sec. 3); Third, what is due him for extra duty in any of the other departments, as when employed by the quartermaster, commissary, or other officer, when extra pay is allowed. (Act Aug. 4, 1854, sec. 6.)

127. The additional pay is entered as follows: First, what is due him for re-enlisting (Act Aug. 4, 1854, sec. 2); Second, what is due him for certificate of merit (Ib. sec. 3); Third, what is due him in lieu of a commission recommended for services in Mexico (Ib. sec. 4). In the columns for bounty due, are entered: First, the three months' extra pay allowed for re-enlisting two months before or within one month after expiration of service (Act July 5, 1838); Second, the instalments of bounty due for re-enlisting at remote stations, together with those that have been paid.

128. Under the head of Clothing is entered all the clothing-money his length of service allows him.

129. " Due the United States" is filled up first, with the stoppages for arms and accoutrements lost or destroyed; Second, the total value of the clothing drawn by the soldier; Third, all other stoppages, as for sentence of court-martial, for damage to public property, for apprehension from desertion, &c. &c.

130. Under "Due the Soldier's Home" is entered the amount of stoppage for the Military Asylum.

131. Opposite the laundress's name is entered the amount due for washing at the time of the soldier's death. Opposite the sutler's name is entered the amount due to him.

132. The dates when the inventories and final statements were forwarded, in obedience to Reg. 152, are entered under the proper headings.

Return of Men Joined.

133. This return is made out for each company, on blanks furnished by the Adjutant-General's Department, and is sent to regimental head-quarters quarterly, in order to furnish a record of all the soldiers in the regiment for regimental head-quarters.

134. The names are entered under the various headings required in the note on the blank, viz.: 1st. *Recruits from Depots;* 2d. *Enlisted in the Regiment;* 3d. *Re-enlisted;* 4th. *By Transfer;* 5th. *From Missing in Action;* 6th. *From Desertion.*

135. Recruits from depots are usually those sent from the general recruiting service to fill up the regiment. Enlisted in the regiment are those who have been recruited by the recruiting officer at the post or station where the company is, or in the vicinity, in contradistinction to those who have been sent from the general recruiting service. Re-enlisted are those whose terms have expired in the regiment and have entered on a new enlistment. By transfer are those who have been transferred by proper authority from some other company or regiment. Missing in action are those who have been dropped from the muster-roll as missing in action, and have subsequently joined again. From desertion are those who have been dropped from the muster-roll as deserters, and have subsequently been apprehended, or have given themselves up. It is not necessary that those men should be present if the officers have been notified officially that the men have been assigned, enlisted, transferred, or have reported themselves at any other post or station, as belonging to the company, they must be taken up. If the missing in action, or desertion, and the return to the company, both occur in the quarter, it is not necessary to note them; in other words, it is not necessary that the soldiers should be dropped and taken up again on the same return.

136. The usual description is entered opposite the name of the soldier; under the heading "Re-enlistment," it is only necessary to state the number of enlistments, and when last discharged. Under the head of Remarks it is usual to state when and how the soldier joined, and, if he is absent from the company, to state where he is.

Quarterly Return of Blanks.

137. This return is required by Par. 2, Order 13, War Department, Feb. 11, 1861, viz. :—

"Every commanding officer of a company will henceforth

7*

PAR. 188.

Form 12.

Quarterly Return of Blanks received and expended in Company "A," 1st U. S. Infantry, in the Quarter ending March 31, 1863.

Date, 1863.	No. of invoice and voucher.	From whom received.	Army Regulations.	Tactics (Casey's).	Bayonet Exercise (McClellan's).	Outpost Duty.	Target Practice.	Muster Rolls.	Muster and Pay Rolls.	Returns, monthly.	Return of men joined.	Reports, morning.	Report book, morning.	Return of deceased soldiers.	Enlistments.	Re-enlistments.	Furloughs.	Descriptive lists.	Discharges.	List for companies.	Final statements.	Certificates of disability.	Remarks.
Jan. 1	1	Lieut. J—— C——, Adjt.	3	3	3	3	3	6	18	12	4	50	1	4	20	20	20	20	15	2	20	20	
March 10	2	Adjutant-General										20			10	10	10	10	10		10	10	
Total to be accounted for			3	3	3	3	3	6	18	12	4	70	1	4	30	30	30	30	25	2	30	30	
Expended								2	4	3	2	40		2	15	15	15	10	10		10	10	Used in making returns.
Lost and destroyed, per certificate			1		1		1		1			5			5	5	5	3	4		4	5	
Total expended			1		1		1	2	5	3	2	45		2	20	20	20	13	14	2	14	15	
Total on hand to be accounted for			2	3	2	3	2	4	13	9	2	25	1	2	10	10	5	7	11	2	16	15	

I certify that the above Return exhibits a correct statement of all the Blanks furnished [...] Quarter ending March 31, 1863.

of Artillery, commanded

	DEATHS.						WOUND	
	Died of wounds received in action.	Accidental.	Missing in action.	Deserted.	AGGREGATE.	In action.		

IRD REGIMENT OF ARTILLEI
cisco, Cal.
1861.

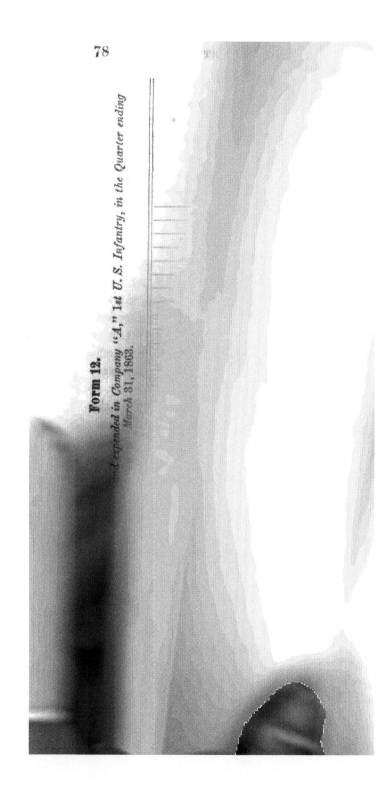

Form 12.

... d expended in Company "A," 1st U. S. Infantry, in the Quarter ending
March 31, 1863.

t
t
d
t
d
ns

om-
that
sary
the
their
and
make
pany

ished.
m 13 is
out this
quired.
calendar

keep a regular account of all books and blanks received and expended by him for the use of his company, and make a quarterly return of the same to the adjutant of his regiment. These returns will be consolidated with those of the regimental head-quarters, and forwarded in this shape by the adjutant, through brigade head-quarters, to those of the division. The assistant adjutant-general at division head-quarters will make similar returns to this office of the books and blanks received by him for distribution to his division."

138. Form 12 is furnished to aid officers in ruling out this return, as no blanks are furnished for the purpose. It shows what blanks companies are likely to receive, and how they are used, and in what proportions. The amount expended is what it is calculated will probably be destroyed in making out and filling up the blanks. (See Requisitions for Blanks, par. 214.)

Annual Return of Casualties.

139. This return is sometimes called for from the companies by the regimental commander, for the reason that the company commanders neglect to make the necessary papers to enable the adjutant to make this return for the regiment. If company commanders have forwarded their monthly company returns, return of deceased soldiers, and men joined, regularly and correctly, the adjutant can make out the annual return without calling upon the company commanders for a return from the companies.

140. No printed forms for company returns are furnished. Blanks are furnished for regimental returns. Form 13 is appended to enable company commanders to make out this return in manuscript on a sheet of letter paper if required.

141. The return is made out at the end of the calendar

year; and to do this it is necessary to have all the records of the company for the year. The company return does not correspond in every particular with the regimental return. The space headed " Designation of Companies, &c." in which the casualties have occurred, is not needed on this return ; simply a statement of the number of deaths, desertions, apprehensions, and surrenders from desertion that have occurred; under " Designation of Posts," a list of the posts and stations at which the deaths, desertions, &c. took place, and the number of each.

142. On the back of the return, the officers' names corresponding with the face of the return are entered under each heading, as *Appointed, Promoted, Transferred,* and *Missing in Action,* under the general heading GAIN. Under the general heading LOSS are entered, under the respective headings, *Resigned, Transferred, Dismissed, Cashiered, Dropped,* and *Died,* with explanatory remarks, as to where from, to what regiment transferred, appointed, promoted, &c., for what dismissed, &c., and place and cause of death. Under the head of Remarks are entered such facts regarding the movements, service, and discipline of the company as are worthy of record. The names of all officers and enlisted men killed or wounded in action, with the *time* and *place,* should also be entered in the column of Remarks, under an appropriate heading.

Certificates of Disability.

143. When a soldier becomes disabled from any cause, as wounds, disease, or infirmity, and is unable to perform his duty, the surgeon notifies his company commander that he considers him a fit subject for discharge, and duplicate *Certificates of Disability* are made out. The printed blanks for this purpose are furnished by the Adjutant-General's

Department. (Form 13, G. R., Med. Dep't.) The first part of the certificate is signed by the company commander, who will add a statement of the facts known to him concerning the-disease, or wounds, or cause of disability of the soldier; the time, place, manner of its occurrence, and all the circumstances under which the injury occurred or the disease originated or appeared; the duty, or service, or situation of the soldier at the time the injury was received, or the disease contracted or supposed to be contracted, and whatever facts may aid a judgment as to the cause, immediate or remote, of the disability. "When the facts are not known to the company commander, the certificate of any officer, or affidavit of other persons, having such knowledge, will be appended."

144. There is also a blank certificate for the surgeon to sign, who fills it out by describing "particularly the disability, wound, or disease; the extent to which it deprives him of the use of any limb or faculty, or affects his health, strength, activity, constitution, or capacity to labor or earn his subsistence. The surgeon will add, from his knowledge of the facts and circumstances, and from the evidence in the case, his professional opinion of the cause or origin of the disability." "When the soldier is a probable case for pension, special care must be taken to state the degree of disability."

145. The certificates, after being signed in duplicate by the captain and surgeon, are sent to department or army head-quarters, where they are submitted to the Medical Director, who signs his approval or disapproval upon them. If disapproved by him, they are returned; if approved, they are submitted to the commanding general, who must endorse on them his order to that effect before the soldier can be discharged. (Reg. 172.)

146. When the certificates are returned with the order

to discharge, the soldier's final statements are made out, and he is discharged. The date of the discharge as given is entered on the certificates, and both are sent to the Adjutant-General at Washington.

147. Regulation 1260 cautions surgeons in giving certificates of disability.

Final Statements.

148. Blanks for Final Statements, according to Form 4, Gen. Reg., Pay Dep't, are also furnished to companies by the Adjutant-General's Department. When a soldier is discharged under any circumstances, except when he forfeits all pay and allowances, he receives, in addition to his *Discharge*, duplicate statements of his pay, clothing account, &c., called FINAL STATEMENTS. They are for the purpose of enabling him to obtain his pay and whatever is due him, from the paymaster, who takes the papers to accompany his accounts when the soldier is paid.

149. The following notes are copied from Form 4, to guide the company commander in these papers, viz.:

Note 1.—The amount of additional pay per month, if any, *for former services*, under the act of *August* 4, 1854, must be carefully noted in the exact words used.

Note 2.—Likewise, the amount due the soldier for a *certificate of merit*, or in lieu of a *commission*, under sec. 4, act of *August* 4, 1854.

Note 3.—So, also, of any other *extra* pay for which he may be mustered; *ex. gr.* as acting *Hospital Steward*, as *Saddler*, &c., and which may be *still due* him.

Note 4.—Stoppages for *loss* or *damage* done to *arms*, or other *public property*, must be noted, and the *articles*, and *particular damage* to each, specified.

Note 5.—Stoppages due under the sentence of a court-martial must be *carefully* stated, with number and date of order.

Note 6.—In cases of *desertion*, the *date*, and that of *delivery* or *apprehension*, must be given, together with a correct transcript of the order of *sentence* or *pardon*, if the man's accounts may be affected by either.

150. The amount of money which the soldier has deposited with the paymaster should also be entered on the final statement. (Reg. 1354, Par. 6.)

For explanation of notes 1 and 2, see Par. 78.

151. Note 3 relates to such extra pay as the paymaster is authorized to pay: the extra pay due from the quartermaster or commissary is not entered on these statements. If those officers have no money to settle with the soldier on his discharge, they must give him a certified account (Form 22, Reg. Q. M. Dep't), adding to the certificate that it had not been paid for want of funds.

Notes 4, 5, and 6.—For explanation of these notes, see Par. 76.

152. In case the soldier loses these statements, officers are prohibited from giving others (G. O. No. 86, 1862). The same order specifies the mode of procedure in such case, viz. :

"Application for payment in these cases must be made through the Paymaster-General of the Army to the Second Comptroller of the Treasury. The application must be accompanied by the soldier's statement, under oath, that his final statements and certificates of discharge are lost, destroyed, or have never been received by him; that he has made diligent search or application for them; that they cannot be recovered or obtained, and that he has not received pay on them, nor assigned them to any other person. All the circumstances of the case must be fully set forth in the affidavit, and this again must be accompanied by all the evidence in corroboration of his statement which the soldier can produce. On receipt of this, the Second Comptroller will audit the accounts, and, if satisfied with the evidence, will order payment to the soldier of the amount found justly due to him.

" The attention of all officers of the army, and particularly

of all company, regimental, and post commanders, surgeons
in charge of general hospitals, and of paymasters, and of
all soldiers discharged from the service, who, from the want
of their final statements and certificates of discharge, are
unable to procure a settlement of their account with the
government, is specially directed to this order." (Reg. 165.)

153. When a soldier dies, his final statements are made out
discharged by death, the same as in any other case, and they
accompany in duplicate the inventory required to be sent
to the Adjutant-General at Washington. If he dies possessed of no effects, that fact is mentioned on the final
statements, and accounts for the inventory not being sent
with them.

154. In computing the retained pay, it must be borne in
mind that the Act of Aug. 3, 1861, sec. 10, made the
amount per month $2, instead of $1 as before; and that
the Act of July 17, 1862, again made it $1.

Discharges.

155. No soldier shall be dismissed from service without
a written discharge, forms for which are supplied by the
Adjutant-General's Department. No discharge is sufficient
unless signed by a field officer of the regiment to which the
soldier belongs, or the commanding officer of the camp or
garrison, when no field officer is present; and no discharge
shall be given before the expiration of the soldier's enlistment, except by order of the President, the Secretary of
War, the commanding officer of a department, or the sentence of a general court-martial. (Art. 11.)

156. Commanding officers of departments have power to
give discharges to soldiers only on a certificate of disability.
This authority is also extended to officers commanding an
army in the field or a *corps d'armée*. After twenty years'

service a soldier may be discharged on his own application. (Reg. 163.)

157. When a soldier's term of enlistment is about to expire, he may be discharged any time within two months before, provided he re-enlists. (Reg. 164.)

158. When a soldier is to be discharged, from any cause whatever, the company or detachment commander fills out the discharge, and, accompanied by the final statements, it is submitted to the officer who is authorized to sign it. The first paragraph of the discharge states where he belongs, and the date and period of his enlistment, to which the cause of his discharge is added, as, "In consequence of the expiration of his enlistment," "On a certificate of disability," or "By sentence of a general court-martial, Orders No. —, dated Head-Quarters Department of the West, St. Louis, Jan. 1, 1863," or "on his application, having served —— years, or —— months," or "By Orders No. —, dated War Dep't, Jan. 1, 1863."

159. The second paragraph of the discharge is a description of the soldier. The part for the character is cut off, unless a recommendation is given; this recommendation may be signed by any officer who knows the soldier. (Reg. 166.)

160. But one discharge is given, and it is in no instance renewed should the soldier lose it. He must in such case resort to some other evidence of service. A certificate of service, or the fact of a discharge having been given, or an affidavit of some one cognizant of the facts, is the only kind of evidence that should be given after a soldier has once lost his discharge. (Reg. 165, Par. 152.)

161. "The giving in duplicate, by any officer of the army, of certificates of discharge or final statements, is peremptorily forbidden (Reg. 165). Not even if such papers are lost or destroyed, is any officer of the army authorized to replace them." (G. O. No. 86, 1862.)

Furloughs, Passes, &c.

162. No soldier can in any case be absent from his proper company without some authority, as an *Order*, a *Furlough*, *Pass*, *Permit*, or something to show that he is either on duty or has permission to be absent.

163. The common evidence is a furlough, or pass: the former is given for long periods, and the latter for periods of one or two days, or a certain number of hours.

The form for a furlough is given in Regulations, page 34. The form for a pass is as follows :—

164.　　　　　FORT SCOTT, KANSAS, Jan. 1, 1862.

Private John Smith, Company "A," 1st U. S. Infantry, has permission to be absent for the purpose of (here state the object of the absence) until Retreat.

　　　　　JOHN BROWN,
　　　　　　　1st Sergt. Co. "A," 1st U. S. Inft'y.
　　　　　A—— B——,
　　　　　　　Capt. Co. "A," 1st U.S. Inft'y.
Approved, J—— D——,
　Col. 1st U. S. Inft'y, Com'dg.

165. The commanding officer of the company or detachment should require the first sergeant to sign the pass first, for the reason that the man may be liable to be detailed for some duty of which the captain is not aware. The pass should be approved by the colonel or commanding officer of the regiment, or by the commanding officer of the post, camp, or garrison.

166. A record should be kept on the *Roster* of *furloughs* and *passes* given, in order that these indulgences may be distributed equally among the soldiers.

167. In time of peace, the 12th Article of War authorizes commanding officers of regiments, posts, or garrisons to give furloughs for such periods as they may think consistent with the good of the service. Captains, or other inferior officers (in the absence of a field officer), may give furloughs for twenty days to two soldiers at a time; but this privilege can only be granted to the same soldiers once in six months. This Article is rather obscure; but the foregoing is believed to be the correct interpretation.

168. In time of war, and when the commanding officer shall deem it necessary, these privileges may be denied. On such occasions, the granting of furloughs, passes, and sick-leaves is regulated by orders published to the troops by commanding officers.

The Act of March 3, 1863, contains the following :—Sec. 32. *And be it further enacted,* That the commanders of regiments and of batteries in the field are hereby authorized and empowered to grant furloughs, for a period not exceeding thirty days at any one time, to five per centum of the non-commissioned officers and privates for good conduct in the line of duty.

Affidavits and Certificates, &c.

169. It frequently becomes necessary to account on the property return for property lost or destroyed, or abandoned, or captured by the enemy. The best evidence that can be obtained should be furnished. This is usually done, First, by obtaining the certificate of some disinterested officer to the facts in the case; Second, the officer's own certificate to the facts, where he is cognizant of them; and, finally, the affidavits of soldiers or citizens. The order to abandon property is sufficient evidence, and throws the responsibility on the officer who issued the order.

170. The usual form of a certificate is as follows :—

I certify, on honor, that a horse, the property of the United States, for which Capt. A—— B——, 1st U. S.

Cavalry, is responsible, died of disease (state the disease if known), at Camp Chase, Ohio, on the 25th of December, 1862.

CAMP CHASE, OHIO, Dec. 25th, 1862.

(Signed duplicates)   C—— D——,

1st *Lieut.* 1st *U. S. Cav.*

171. The same form is used for any other fact concerning property lost, or destroyed, or abandoned.

172. The form of an affidavit is that usually used in civil courts, as follows :—

Personally appeared before me, Private A—— B——, Company A, First U. S. Cavalry, who, being duly sworn, says, that whilst on the march in pursuit of the enemy, on the Rappahannock River, on the 15th of December, his horse, the property of the United States, for which Captain A—— B——, First U. S. Cavalry, is responsible, became so worn out and exhausted that he was compelled to abandon him.

Sworn to and subscribed before me, this 20th day of Dec'r, 1862. } A—— B——, *Private Co. A, First Cavalry.*

J—— C——,

First Lieut. and Adjutant First Cavalry.

(Signed duplicates.)

173. Affidavits are usually taken before the following officers, in the order given :—1st. By a civil magistrate, competent to administer oaths; 2d. A judge-advocate. 3d. Recorder of a garrison or regimental court-martial; 4th. Adjutant of the regiment; 5th. A commissioned officer. (Reg. 1031.)

174. An officer may make his own certificate, as above, or he may consolidate the articles on an abstract, Form

8, and certify at the bottom to the correctness of the remarks, opposite.

The Act of Feb. 7, 1863, sec. 2, authorizes that a company commander may account for the loss of clothing or military supplies, or any apparent deficiency thereof, by his own affidavit, setting forth the loss of vouchers or company books, or any matter or circumstance tending to show that the loss or apparent deficiency was the result of unavoidable accident or actual service, without any fault on his part; and that the articles were properly and legally used and appropriated.

Inventories of Deceased Soldiers.

175. The regulations with regard to the effects of deceased soldiers are derived from the 95th Article of War, Reg. 152 to 154. When a soldier dies, the commanding officer of the company or detachment is required to take an account of the effects of which the soldier died possessed. This account, called the inventory, is made out according to Form 14, in triplicate; one copy is sent to the Adjutant-General of the army, one to the regimental head-quarters, and one is retained by the officer. Duplicate Final Statements accompany this inventory.

176. When a soldier dies away from his company, these papers are made out, if he dies in a general hospital, by the surgeon, otherwise, by the commanding officer of the detachment, and forwarded as above. He also forwards to the commanding officer of the company to which he belongs a report of the soldier's death, specifying the *date, place, and cause,* to what time he was *last paid,* and the *money* or other *effects* of which he died possessed. The company commander is thus enabled to account properly for the soldier on the next Muster Roll. Should the soldier have no effects, it will be so stated on his Final Statement.

PAR. 175. **Form 14.**

Inventory of the effects of —— late a —— of Captain —— Company, ( ) of the —— Regiment of United States —— was enlisted by —— of the —— Regiment of —— at —— on the —— day of —— 186 , to serve —— years ; he was born in —— in the State of —— is —— years of age, —— feet —— inches high, —— complexion, —— eyes, —— hair, and by occupation when enlisted a ——; he died in —— at —— on the —— day of —— 186 , by reason of ——.

INVENTORY.

ARTICLES.	No.	ARTICLES.	No.
Hats..............................	*	Cotton shirts..................	
Caps..............................		Pairs boots....................	
Forage caps.....................		Pairs shoes....................	
Great coats.....................		Pairs socks....................	
Uniform coats..................		Blankets......................	
Uniform jackets...............		Haversacks....................	
Flannel sack coats...........		Knapsacks.....................	
Blouses..........................			
Stable frocks..................			
Fatigue overalls..............		MONEY. Specie........... $	
Pairs trowsers,...............			
Pairs flannel drawers,......			
Pairs cotton drawers........		Notes............. $	
Flannel shirts..................			

I certify, on honor, that the above inventory comprises all the effects of —— deceased, and that the effects are in the hands of —— at —— to be disposed of by a Council of Administration.

(Duplicates.)

Station : ——
Date : ——

——— ———,

——— ———,
Commanding the Company.

NOTE 1.—The effects in all cases, when called for, should be turned over to the legal representative, without further authority from the Adjutant-General. When the effects are turned over to the relatives of the deceased before these inventories are sent to the Adjutant-General, their receipts therefor should be attached to the inventories. In all other cases, one copy will be sent with a letter of transmittal direct to the Adjutant-General, and a duplicate retained by the officer.

NOTE 2.—Particular care should be taken to take receipts in duplicate from the Paymaster for any funds turned over to him; one copy should be sent direct to the Adjutant-General, and one retained by the officer.

These Final Statements will materially facilitate the settlement of the soldier's account on the application of heirs for any thing that may be due the deceased from the Government. If the heirs have received the soldiers's effects, it will be stated on the inventory.

177. Should no heirs or administrators appear to take charge of the soldier's effects within a short period of his decease, a council of administration is required to dispose of them, under the direction of the commanding officer. This is usually done by the officer in charge of the effects on the order of the council, approved by the commanding officer. Public notice is given throughout the camp or garrison of the articles to be sold, and of the day and hour of sale: they are then sold at auction to the highest bidder. The proceeds are turned over to the nearest paymaster, and kept to the credit of the United States until claimed by the heirs or legal representatives. An account of the proceeds is certified to by the council and commanding officer, and, accompanied by the paymaster's receipt, it is sent by the commanding officer to the Adjutant-General.

178. The Inventory will be endorsed, "Inventory of the effects of —— ——, late of Company —, —— Regiment of ————, who died at —— the — day of ——, 186—."

179. The Report will be endorsed, "Report of the proceeds of the effects of —— ——, late of Company —, —— Regiment of ————, who died at —— the — day of ——, 186—."

180. The Inventory is made out immediately after the soldier's death, and should be forwarded as soon as possible. The Report is sent as soon after the sale as possible. A record of the dates when they are sent should be kept, as it must be given on the Return of Deceased Soldiers.

Company Council of Administration.

181. Every two months at least, and whenever it is necessary, the company commander convenes all the officers for the purpose of making appropriations from the company fund for the benefit of the company. The company council is also convened every four months, also whenever another officer takes command of the company, and when the company leaves the post, for the purpose of auditing the company fund.

182. The council is usually convened by an order issued by the company commander. The company fund is audited only at the times above specified : at the other meetings, the council only meet for the purpose of making appropriations. The company commander has possession of the fund, and generally disburses the money when necessary, and presents the bills to the council to obtain its approval and the necessary appropriations. The proceedings are entered in the account-book of the company fund in the following form, and signed by all the officers of the council, viz. :—

183. *Proceedings of a Company Council of Administration, convened pursuant to the following Order, viz. :—*

CAMP CHASE, OHIO, April 29, 1863.

ORDERS No. 4.

A Company Council of Administration will convene tomorrow, at 10 o'clock A. M., (at the company orderly room), or as soon thereafter as practicable, for the purpose of auditing the Company Fund.

A—— B——,

Capt. First U. S. Infantry,

Commanding Company A.

CAMP CHASE, OHIO, April 30, 1863.

Council met, pursuant to the above order. Present,

Capt. A—— B——, First Lieut. D—— C——, and Second Lieut. J—— D——, Company A, First U. S. Infantry.

The Council then proceeded to audit the Company Fund.

Balance on hand, last Return..............................	$ 50 00
Received from Post (or Regimental Fund)..............	170 00
Received from sales Company Savings, January, February, March, and April	327 00
Total...	$547 00
Appropriated by order of the Council, January, February, and March...	$225 00
Balance on hand, April 1..................................	$322 00

The Council then made the following appropriations for the month of April, viz. :—

For four Brooms, 25 cents each..............................	$ 1 00
Two barrels Onions, $5 per bbl..............................	10 00
Twenty bushels of Potatoes, 40 cents per bu............	8 00
Ten lbs. Pepper, @ 20 cts. per lb..........................	2 00
One sett of Carpenter's Tools................................	15 00
One Cooking Range...	40 00
Books for Company Library..................................	25 00
	$101 00
Balance to be carried to next account.......................	$221 00

There being no further business before it, the Council adjourned. '

A—— B——,

Capt. Comd'g Co. A, 1st U. S. Inf'ty, President.

D—— C——,

First Lieut. 1st U. S. Inf'ty.

J—— D——,

Second Lieut. 1st U. S. Infantry, Recorder.

184. The book containing the record of these proceedings is sent in to the head-quarters of the post or regiment, and the approval of the commanding officer obtained.

Provision Returns.

185. Subsistence for enlisted men is obtained upon Provision Returns, according to Form 15 (see Form 14, Subsistence Reg.), which is here inserted for convenience, as it is a' form constantly required, and printed blanks are not always to be had. Form 16 is inserted to aid in ascertaining the bulk of any number of rations. This return should be made by the commanding officer of the company or detachment. In a regiment all the company returns are consolidated in the adjutant's office, and at a post all the returns for the different companies and detachments are also thus consolidated, and the provisions drawn upon the consolidated return by each company and detachment separately. The order for the issue is signed by the commanding officer of the regiment or post.

186. When all the provisions are not desired by the company, those not wanted are left at the commissary store, and a memorandum receipt taken for them from the commissary. At the end of the month these receipts are consolidated on one account, Form 17, and the commissary buys them. This account is made out in triplicate : one is retained by the commissary, and the other two by the company, until it can be paid. The money thus obtained goes to the company fund. (Par. 39.)

187. When rations are received by the company which are unfit for issue, from any cause, the company commander applies to the commanding officer for a Board of Survey. Should the board condemn the provisions, a special return is made of the provisions condemned, and they are replaced. The proceedings of the board accompany the return, to be filed with the commissary's vouchers.

...ions

No. of Rations.	Barrels.		Quarts.
1	....	..	.....
2	....		....
3	....		....
4	....		....
5	....		.....
6	....		.....
7	....		.....
8	....		.....
9	....		.....
10	....		.....
20	....		....
30	....		1
40	....		1
50	....		2
60	....		2
70	....		2
80	....		3
90	....		3
100	....		.....
1,000	3	1	......
10,000	37	1	......
100,000	375	...	......

NOTES.—Fresh potato shel ...
or peas; m...
* Roasted or a rat...
See par. 21, p...

Par. 185.

Provision Return of —— Company, —— Regiment of ——, for —— days, commencing the —— day of ——, 186 , and ending the —— day of ——, 186 .

Form 15.

STATION.	Number of men.	Number of women.	Total.	Number of days.	Number of rations.	RATIONS OF																		REMARKS.
						Pork.	Salt beef.	Fresh beef.	Flour.	Beans.	Rice.	Coffee.	Tea.	Sugar.	Vinegar.	Sperm candles.	Adamant. candles.	Soap.	Salt.					

The A. C. S. will issue on the above return,

————, *Commanding Officer.*

————, *Commanding Company.*

NOTES.—This return must embrace only the *actual* strength of the company *present*, including the authorized company women *present.*
The sick in hospital, hospital attendants, &c., will be returned for, by the medical officer in charge, on a *separate* return.
Subsistence stores for the use of officers, their families, and authorized servants, may be *purchased* from the commissariat.
Provisions should ordinarily be drawn for a period of from five to ten days at a time.

PAR. 186.

FORM 17.

THE UNITED STATES, To ———, Comd'g ——— Company, ——— Reg't ———,   DR.

186		DOLLARS.	CENTS.
	$$		

I certify that the articles above mentioned are the actual savings of my company for the month of ———, 186 .

———, *Comd'g Comp'y.*

Examined:

———, *Comd'g Officer.*

I certify that the above account is correct and just; that the articles mentioned have been taken up on my "Return of Provisions" for the month of ———, 186 , and that I have not paid the account.

———, *A. C. S.*

Received at ———, this —— day of ———, 186 , from ———, A. C. S., U.S.A., —— dollars and —— cents, in full of the above account.

———, *Comd'g Comp'y.*

NOTE.—This paper is to be made out and signed in triplicate—one copy for the Commissary who takes up the savings (to accompany his "Return of Provisions" *for the month*), and two copies for the Commissary who pays the account.

188. Should the company lose their rations, or any portion of them, from any unavoidable cause or accident, a special return is made for the provisions lost, which is accompanied by the certificate of the officer cognizant of the facts, stating the cause, and whether through fault of any one in the military service. In the absence of a commissioned officer, the affidavit of an enlisted man or citizen accompanies the return and becomes a voucher for the commissary. Should the provisions be lost or destroyed through the fault of any one in the military service, they may be charged against him on his pay-roll or account.

189. Only the soldiers actually present are drawn for on the provision returns. Soldiers absent with authority, sick in hospital, or on furlough, are not drawn for.

190. The ration during the present insurrection is as follows, viz. :

MEAT........	$\frac{3}{4}$ lb. of Pork or Bacon, or
	1$\frac{1}{4}$ lb. of Fresh or Salt Beef.
BREAD.......	22 oz. Soft Bread or Flour, or
	1 lb. Hard Bread, or
	1$\frac{1}{4}$ lb. of Corn Meal.
BEANS......	8 qts. to 100 rations of Beans or Peas.
RICE........	10 lbs. to 100 rations of Rice or Hominy.
COFFEE.....	8 lbs. to 100 rations of Roasted or Ground Coffee, or
	10 lbs. to 100 rations of Green Coffee, or
	1$\frac{1}{4}$ lb. to 100 rations of Tea.
SUGAR......	15 lbs. to 100 rations of Sugar.
VINEGAR...	4 qts. to 100 rations of Vinegar.
CANDLES...	1 lb. to 100 rations of Sperm Candles.
	1$\frac{1}{4}$ lb. to 100 rations of Adamantine Candles.
	1$\frac{1}{2}$ lb. to 100 rations of Tallow Candles.
SOAP........	4 lbs. to 100 rations of Soap.
SALT........	2 qts. to 100 rations of Salt.
MOLASSES.	1 gal. to 100 rations, or $\frac{1}{100}$ gal. to each man, twice in seven days.
POTATOES..	1 lb. to each soldier, three times in seven days.

191. When beans, peas, rice, hominy, or potatoes cannot be issued in the proportions above given, an equivalent in *value* shall be issued in some other proper kind of food.

192. Desiccated potatoes, or mixed desiccated vegetables, may be substituted, at the rate of an ounce and a half of potatoes or one ounce of mixed vegetables to the ration, for beans, rice, or fresh potatoes, or their equivalent of peas, hominy, &c., when these articles cannot be had, and when the supply on hand will admit of it, twice per week, on the approval of the commanding officer.

193. Fresh beef may be issued as often as the commanding officer of any detachment or regiment may require, if practicable, instead of salt meats.

194. *Extra* issues are allowed only as follows, viz. :—ten pounds of sperm, or twelve pounds of adamantine, or fifteen pounds of tallow *candles* to the principal guard of each camp or garrison, on the order of the commanding officer, for which a return is made as for other issues of provisions, signed by the commanding officer. This return is made by the adjutant, and the candles are issued daily in proportion to the allowance. *Salt,* not exceeding two gills a month to each public animal, the return to be made by the officer accountable for the animals, and the order for the issue signed by the commanding officer. *Whiskey,* one gill daily to each man, in case of "excessive fatigue or exposure."

195. Without further legislation by Congress, at the close of the present insurrection the ration will be as follows, viz. :—

MEAT ..... { ¾ lb. of Pork or Bacon per ration, or
1¼ lb. of Fresh Beef per ration.

BREAD...... { 18 oz. of Bread or Flour, or
12 oz. of Hard Bread,
1 lb. of Hard Bread, in the field or on transports, } per ration.

BEANS......   8 qts. of Beans or Peas, or } per 100 rations.
              10 lbs. of Rice,
COFFEE....   10 lbs. of Coffee per 100 rations.
SUGAR......   15 lbs. of Sugar per 100 rations.
VINEGAR...   4 qts. of Vinegar per 100 rations.
CANDLES..   1 lb. of Sperm, or
             1¼ lb. of Adamantine, or } per 100 rations.
             1½ lb. of Tallow Candles,
SOAP.......   4 lbs. of Soap per 100 rations.
SALT........   2 qts. of Salt per 100 rations.

196. Laundresses are returned for with the companies to which they belong, but are often permitted to accumulate their rations in the commissary store to the end of the month. Laundresses who are absent from the companies to which they belong temporarily, should have a certificate from the company commander that they are the laundresses of his company. A laundress can only draw rations in kind, and only when she is at a post or camp where they can be issued.

197. Soldiers on furlough by proper authority can have their rations commuted at the cost of the ration at the post where they belong : they are therefore not returned for with the company during their furlough.

Requisitions.

198. Requisitions are forms used for specifying the requirements of troops for certain authorized allowances for their support and service, to which the order for the issue is obtained from the proper authority. They are necessary for placing before the proper officer, in an official manner, the articles wanted and the necessity for their issue. Requisitions are of two kinds, *regular* and *special*. A regular requisition is for such articles where the allowance is regulated and fixed by law or regulation. A special requisition is where the articles required are rendered

necessary from some cause, and for articles for which there
is no regulated allowance, for the benefit of the public service.   Printed blanks are generally furnished by the
department from which the property is obtained.

199. The following requisitions are the most used, viz. :—

REQUISITION FOR FUEL.—Fuel is furnished by the
quartermaster, upon the requisition of the officer requiring
it, either for himself, or for the company or detachment he
commands, and only for the month in which it is due. The
allowance of fuel for companies is as follows, viz. :—

	Cords of wood per month.	
	From May 1 to Sept. 30.	From Oct. 1 to April 30.
Captain..	¾	3
Lieutenant.......................................	½	2
Each enlisted man, officer's servant, or laundress	¹⁄₁₂	⅛
Wagon and Forage Master, Sergeant-Major, Quartermaster-Sergeant, and Ordnance-Sergeant..........................	½	1
Each guard-fire regulated by the commanding officer.................................	......	3

200. Merchantable hard wood is the standard.   The
cord is 128 cubic feet.   Coal may be substituted for wood
at the rate of 1500 pounds of anthracite or 30 bushels of
bituminous to the cord.   Two cords of soft wood, such as
pine, in lieu of one of hard, may be issued at the discretion
of the department commander.

201. Fuel is Government property, and when the allowance is not used by officers or troops the surplus is
returned to the quartermaster, who is required to take it
up on his return as surplus.   The fuel for baking bread

may be taken from the surplus, if there is any, but the balance, if any, must be turned in.

202. In November, December, January, and February, the foregoing allowance is increased one-fourth north of the 39th degree of latitude, and one-third north of the 43d degree of latitude.

203. REQUISITION FOR FORAGE.—Forage is also furnished by the quartermaster, and only in the month for which it is due. Fourteen pounds of hay, and twelve pounds of oats, corn, or barley, is the allowance for horses; fourteen pounds of hay, and nine pounds of oats, corn, or barley, for mules.

204. Forage, like fuel, is not to be drawn and sold: it must be used, or turned in as Government property, and accounted for. Commutation for forage is only allowed to officers, and only when it cannot be furnished in kind, under which circumstances a certificate is necessary from the quartermaster whose duty it was to furnish it, to the effect that it could not be furnished, and the reason therefor. When officers are detached, and there is no quartermaster, the officer desiring commutation must have the certificate of some other officer cognizant to the facts; and, if there is no other officer, the officer must make his own certificate, and state in addition that he was detached, that there was no quartermaster or other officer, and that he has not drawn any forage for the period stated.

205. When the exigencies of the service make it necessary, the forage-ration may be reduced by the commanding officer, who will prescribe the allowance of forage under such circumstances.

206. STRAW.—Requisitions for straw are also filled by the quartermaster. The allowance for each enlisted man, officer's servant, and laundress is twelve pounds per month, allowed in barracks only. One hundred pounds per month is allowed for each horse in public service. Where posts are near public lands where hay can be cut by the troops, it will be so provided in lieu of straw. Straw not used is to be accounted for as other surplus property.

207. In barracks, company commanders should draw bed-sacks for the use of the soldiers, and the allowance above will keep the men amply supplied.

208. STATIONERY.—Requisitions for stationery are made quarterly, and are filled by the quartermaster. The allowance for a company and its officers is as follows, viz. :—

	Quires of writing-paper.	Quires of envelope-paper.	Number of quills.	Ounces of wafers.	Ounces of sealing-wax.	Papers of ink-powder.	Pieces of office tape.
Commanding officer of a company..............	5	$\frac{1}{2}$	20	$\frac{1}{2}$	8	1.	1
Lieutenant not commanding company........	$1\frac{1}{2}$	$\frac{1}{8}$	6	$\frac{1}{8}$	1	$\frac{1}{4}$	$\frac{1}{4}$

209. Steel pens, in the proportion of one holder to twelve pens, may be issued in lieu of quills; and one hundred envelopes to one quire of envelope-paper.

210. When an officer is relieved from the company, he should not take any of the stationery with him which he drew as commanding company.

211. Blanks and blank-books furnished by the Quartermaster's Department are classed under the head of stationery. For the company the following are obtained on special requisition, when needed for the use of the company, viz. :—

Morning Report Book.
Descriptive Book.
Clothing Book.

Order Book.

Record Book for Target Practice.

212. Other blank-books required are either purchased with the company fund or made out of the stationery furnished.

The blanks furnished by the Quartermaster's Department are,—

Returns of Clothing, Camp and Garrison Equipage.

Receipt Rolls of Clothing.

Invoices and Receipts of Clothing.

Returns of Quartermaster's Property.

Invoices and Receipts of Quartermaster's Stores.

213. Small deficiencies of the foregoing books and blanks may be obtained by requisitions upon the nearest quartermaster.

214. Ordinarily, however, blank-books and blanks of all descriptions are obtained through regimental headquarters. The commanding officer of the regiment makes a requisition upon the adjutant-general of the division or department for a six months' supply, upon the receipt of which they are distributed *pro rata* to the companies. The following is regarded as a six months' supply for a regiment of ten companies, viz. :—

1 Guard Report Book.

1 Consolidated Morning Report Book.

10 Company Morning Report Books.

100 Consolidated Morning Reports.

2 Lists of Rolls and Returns for Companies.

6 Field and Staff Muster Rolls.

60 Monthly Returns.

6 Quarterly Returns of Deceased Soldiers.

2 Annual Returns of Casualties.

100 Non-Commissioned Officers' Warrants.

6 Hospital Muster Rolls.

18 Field and Staff Muster and Pay Rolls.

18 Hospital Muster and Pay Rolls.

60 Company Muster Rolls.

180 Company Muster and Pay Rolls.

12 Regimental Returns.

20 Returns of Men Joined.

30 Quarterly Company Returns of Deceased Soldiers.

40 Descriptive Lists.

—— Certificates of Disability.

—— Blank Discharges.

—— Final Statements.

—— Furloughs.

—— Enlistments.

—— Re-enlistments.

—— Regimental Recruiting Returns.

These blanks are furnished or required for in proportion to the wants of the regiment.

215. Where the troops are brigaded, all requisitions for books and blanks furnished by the Adjutant-General's Department must be made through regimental and brigade head-quarters, in each division, to the assistant adjutant-general at division head-quarters, who will himself, from time to time, make general requisitions on the Adjutant-General at Washington for the supply of his division.

216. Where troops are not brigaded, as sometimes happens with artillery, requisitions for books and blanks will be made upon the chiefs of their respective arms in the army in which they are attached, and returns will be made to these officers as though to division head-quarters; and the chiefs of artillery and cavalry will make requisitions for the books and blanks needed for their commands, and make returns of the same, to the Adjutant-General's office, as prescribed in Par. II., General Orders No. 13, 1862. (See Par. 137 and 138.)

217. A regiment newly organizing will be allowed the following books, viz. :—

35 Regulations,	30 Target Practice,
35 Tactics,	35 Outpost Duty.

30 Bayonet Exercise.

218. This is for a regiment of ten companies. For a regiment of twelve or twenty-four companies the allowance is in proportion,—the principle being to furnish each officer with a copy of each of the foregoing books. When an officer is newly appointed, and his regiment has been supplied, he may get the foregoing by letter to the adjutant-general of division, or, if there is none, to the Adjûtant-General of the army direct, at Washington.

Requisitions for Arms, &c.

219. Requisitions for arms, accoutrements, horse-equipments, ammunition, &c. are usually made by commanders of posts and regiments for their entire commands, and distributed to the companies. To ascertain the amount required, the regimental or post commander calls for requisitions from the various companies and detachments, which are then consolidated, and the consolidated requisition is sent up through the intermediate commanders to the Ordnance Department, if they cannot be obtained before. Department, division, or corps commanders frequently have the means of supplying limited deficiencies.

220. A regiment or company once completely armed and equipped cannot be re-supplied until the arms and equipments already issued to them have been inspected and condemned or ordered to be turned in. The requisition must pass through the Ordnance Department, except in cases of necessity and when the department or corps commanders have the means of supplying them. Ammunition can ordinarily be supplied without the approval of the Chief of Ordnance, as ordnance officers are generally located convenient for the supply of troops. Arms, however, are not so readily obtained, except where they have accumulated from troops in the command that have ceased to require them and have turned them in to the nearest ordnance office.

221. Printed blanks for requisitions for ordnance may be obtained from any ordnance officer, according to Form 18. (See Form 22, Reg., Ord. Dept.)

Requisitions for Clothing, Camp and Garrison Equipage.

222. Regimental or post quartermasters draw the supplies from depots, and company commanders make their requisitions upon them through the regimental or post commanders who order the issues.

223. For a regiment in the field, it is usual to make periodical issues, and for the commander of the regiment to call for requisitions from the various company commanders, who should ascertain, by actual inspection, what is required. These requisitions are then consolidated, and the regimental quartermaster draws from the depot for the entire regiment.

224. Company commanders should keep no clothing on hand, for their own security, as well as because it is prohibited.

225. Clothing may be obtained according to the wants of the troops; but the amount of camp and garrison equipage is limited. The following is the allowance for a company, viz. :—

	Wall tent.	Sibley tent.	Common tent.	Spades.	Axes.	Pickaxes.	Hatchets.	Camp-kettles.	Mess pans.	
Captains...............	1	...	...	...	1	...	1	...	...	
Subalterns, to every two.................	1	...	...	...	1	...	1	...	...	
To every 20 foot or 17 mounted men	...	1	...	2	2	2	2	2	5	Or one common tent to 6 foot or 5 mtd. men.

PAR. 221.

Special Req

Where station

Required to arm the whole Company, Regiment, Fort, or Brigade.

I certify th

Examined

The above

(To face

226. In addition to the foregoing, when in garrison, bed-sacks are allowed, and iron pots instead of camp-kettles. One guidon and two trumpets or bugles are allowed to each mounted company.

227. Camp and garrison equipage includes the cap and hat ornaments, musical instruments, and haversacks, canteens, and talmas, &c. (See note to the Price List of Clothing, Camp and Garrison Equipage.) They are borne upon the property return, and are only charged to the soldier when lost or destroyed by him.

228. Other articles of camp and garrison equipage allowed to the various arms of service are stated on the List of Camp and Garrison Equipage. The allowance, however, is not fixed by any existing orders or regulations.

229. The officer who receipts for the clothing, camp and garrison equipage, must account for it, and is held responsible for it until he turns it over to some other officer. (See Return of Clothing, Camp and Garrison Equipage, Par. 94.)

230. During active campaign the allowances above mentioned are materially reduced. No surplus clothing and no tents are allowed, except shelter-tents. The latter are allowed at the rate of one shelter-tent to each commissioned officer, and one to every two non-commissioned officers, soldiers, officers' servants, and authorized camp-followers.

Requisitions for Quartermaster's Property.

231. Company commanders of infantry are rarely responsible for quartermaster's property. The regimental quartermaster usually is responsible for the authorized teams and wagons for the regiment.

232. Artillery and mounted companies have usually the following quartermaster's property to account for, viz.: horses, mules, wagons, harnesses, picket-lines, &c.

PAR. 233.

Form 19.

Special Requisition.

For _____

I certify that the above requisition is correct; and that the articles specified are absolutely requisite for the public service, rendered so by the following circumstances:

....

....

....

....

_____Quartermaster U. S. Army, will issue the articles specified in the above requisition.

_____

Commanding.

Received, at _____, the_____ of_____ 186

of_____ Quartermaster U. S. Army,

in full of the above Requisition. (Signed in triplicate.)

PAR. 286.

Form 20.

Inventory and Inspection Report of ———— for which ———— Regiment ———— is responsible, and which has been inspected and reported on by ————.

		INVENTORY.				INSPECTION REPORT.	
Articles.	No.	Quantity and Quality.	How long in use.	How long in possession.	From whom received.	Condition.	Disposition.

The above is a correct inventory of Quartermaster's property for which I am responsible, and which, in my opinion, requires the action of an Inspector.

(To be made in triplicate.)

I certify that I have examined each article set forth in the Inventory hereto attached, and its condition is as stated in the remarks opposite to it in the above Inspection Report.

STATION : ....

DATE : ....

10

233. These are obtained through the regimental and post quartermasters on requisitions, as any of the foregoing. A form is inserted for making requisitions, which is used in all cases where there is no other prescribed form. (Form 19.)

234. It must be borne in mind that all property receipted for, except fuel, forage, straw, stationery, and rations, which are regular and daily allowances, must be taken up on a property return and accounted for at the end of the quarter in which it is received. Invoices corresponding to the receipts should be taken: they are necessary as a memorandum and official statement of what has been received.

Inventories and Inspection Reports.

235. Officers are prohibited from dropping from their returns any property as *worn out*, or *unserviceable*, until it has been properly inspected and ordered to be dropped. Therefore, when public property in the hands of an officer becomes worn out or unserviceable, he applies for an inspector through his immediate commander. This is usually done by a letter addressed to the authority that has power to order the inspection, which is sent up through the intermediate commanders. At separate posts the commanding officer is authorized to be the inspector. In armies in the field, the commanding officer, or a corps or division commander, orders the inspection.

236. The inventories are made out according to Form 20. The inspection report is left blank for the inspector to fill up. They are made out in triplicate, and the different kinds of property, as ordnance, clothing, camp and garrison equipage, and quartermaster's property, should be kept separate on different inventories. It is also best to separate the property according to its probable disposition: thus, what is to be dropped as entirely worthless, and what is to be turned in as unserviceable, but repairable or to be used for some other purpose. These papers are filled up by the inspector and sent by him to the authority ordering the inspection, for his orders in the case. If the inspection is made by the commanding officer of the post, the reports are sent to the department commander for his orders. Two of the reports are sent back to the officer accountable for the property, and become his vouchers for the disposition of the property; the other is sent by the corps or department commander to the Bureau to which the property belongs. Of the two sent to the officer, one is sent with the returns of the property for which the officer is accountable, and the other is retained.

237. These inspections must not be confounded with the action of Boards of Survey. The former is only intended for the *condemnation* of public property, and the latter is for assessing the damage to public property and fixing the responsibility, and also for fixing the reduced rates at which damaged clothing, from any cause, may be issued to soldiers.

Board of Survey.

238. Boards of Survey are called in the following cases, viz.: To ascertain and to assess the damage to public property from any extraordinary cause, other than fair wear and tear, and to determine who is responsible, whether the carrier, or the person immediately in charge of the property, or the officer who has the property on his returns.

239. To examine and report upon the loss or deficiency of public property, and to fix the responsibility of such loss or deficiency.

240. To make inventories of property ordered to be abandoned or broken up, when the articles are not enumerated in the order.

241. To assess the price at which damaged clothing may be issued to troops, and to fix the proportion in which damaged supplies may be issued to make them equal to the full allowance.

242. To ascertain the discrepancy between invoices and the property actually received, and to fix the responsibility of the discrepancy, whether with the officer who sent them, the carrier, or the contractor.

243. To make inventories and report the condition of the property for which an officer who has died is responsible at the time of his death.

244. When any officer or soldier shall lose or damage by neglect or fault any article of public property, the Board of Survey shall assess the value, damage, or cost of repair of such articles, which the officer or soldier shall pay for the same.

245. Boards of Survey are ordered by commanding officers by their direction, or on application of an officer, and consist usually of three officers, from which the commanding officer and the officer responsible for the property are excluded, except when these two only are present, and then the one not accountable for the property will act; and the responsible officer will constitute the Board only when there is no other resource.

246. The proceedings of the Board are sent in to the commanding officer, and are complete on his approval. They are signed by each member, and a copy is forwarded to department head-quarters or the head-quarters of the corps commander, by the approving officer, and two copies to the officer accountable for the property.

247. The necessity for Boards of Survey on company property usually arise when clothing is received; the original packages may be deficient, or the clothing may be damaged. It is then customary to address a letter to the adjutant, asking for a Board of Survey, and stating the property for which the Board is called to investigate.

Inventories for Boards of Survey.

248. When the Board meets, the officer responsible presents the property, with an inventory, stating the condition of the articles. This inventory is nothing more than a list of the articles, with remarks opposite, setting forth their condition and cause of damage or deficiency. This list is embodied in the proceedings of the Board.

FORM FOR A BOARD OF SURVEY.

249. Proceedings of a Board of Survey, convened at Camp Chase, in obedience to the following order, viz. :—

> HEAD-QUARTERS 2D U. S. CAVALRY,
> CAMP CHASE, OHIO, Jan. 10, 1863.

ORDERS No. 6.

A Board of Survey, to consist of Capt. A—— B——, 1st Lieut. D—— C——, and 2d Lieut. A—— J——, 2d U. S. Cavalry, will convene to-day at 11 o'clock A.M., or as soon thereafter as practicable, for the purpose of investigating and reporting upon the deficiency of certain articles of clothing for which Capt. L—— M——, 2d U. S. Cavalry, is responsible.

> By command of Col. K——.
> (Signed)   J—— J——,
> 1st *Lieut. and Adjt. 2d U. S. Cavalry.*

> CAMP CHASE, OHIO, Jan. 10, 1863.

The Board met pursuant to the above order. Present, Capt. A—— B——, 1st Lieut. D—— C——, and 2d Lieut. A—— J——, 2d U. S. Cavalry.

The Board then proceeded to inquire into the deficiency of the following articles of clothing, viz. :—

Five Blankets,
Three Great-Coats,
Five pairs Bootees,
Ten pairs of Stockings.

First Sergt. J—— C——, Company A, 2d U. S. Cavalry, being duly sworn, says, that he opened the original packages of clothing, and that in two packages of blankets, marked ——, there were three blankets missing in No. —— and two in No. —— In one package of great-coats, marked —— and numbered ——, there were three great-coats missing. In one box of bootees, marked —— and numbered ——, there were five pairs deficient. In one box of stockings, marked —— and numbered ——, there were ten pairs missing. The packages did not seem to have been opened before, and were in good order and condition when received.

Quartermaster-Sergeant B—— C——, 2d U. S. Cavalry, being duly sworn, says:—I was present when Sergeant C —— opened the packages, and know that the facts to which he has testified concerning the foregoing deficiencies are correct.

The Board verified the marks and numbers of the packages, and are of the opinion that the deficiency was in the original packages, and that they were not correctly marked at the time they were packed, and that Capt. L—— M ——, 2d U. S. Cavalry, is in no way responsible for the deficiency.

There being no further business before the Board, it adjourned *sine die.*

A—— B——,

Capt. 2d U. S. Cavalry, President.

D —— C——,

1st Lieut. 2d U. S. Cavalry.

A—— J——,

2d Lieut. 2d U. S. Cavalry.

Recorder.

Approved,

J —— K——,

Col. 2d U. S. Cavalry,

Commanding.

250. Boards of Survey should not omit to take the depositions required by law in the case of clothing. (See Par. 1168, Reg., and Par. 97.) Also with regard to other property, where the evidence is derived from a soldier or citizen

Par. 251.

Form 21.

Return of Killed, Wounded, and Missing in Company A, 1st U. S. Infantry, at Bull Run, Va., July 21, 1861.

NAMES.	Rank.	Killed.	Mortally.	Severely.	Slightly.	Missing.	REMARKS.
			WOUNDED.				
J—— D........	Captain.	1					Shot through the head.
C—— L........	2d Lieut.				1		Flesh-wound in the left arm.
A—— M........	Sergeant.		1				Shot through the lungs.
J—— F........	Corporal.	1					Shot through the heart.
P—— C........	Private.	1					Both legs carried away by a cannon-ball.
J—— N........	"		1				Shot through the abdomen.
C—— F........	"			1			Leg fractured by a shell.
K—— L........	"				1		Grazed by a fragment of a shell.
T—— D........	"					1	Supposed to be wounded and a prisoner.
L—— M........	"					1	Supposed to be captured.
O—— P........	"					1	"
		3	2	1	2	3	

I certify that the above is a correct return of the killed, wounded, and missing in Company A, 1st U. S. Infantry, at the engagement at Bull Run, Va., July 21, 1861.

C—— F——,
1st Lieutenant 1st U. S. Infantry,
Commanding Company.

PAR. 254.

Form 22.

Report of Target Practice to ascertain the Best Shot in the Companies of the 1st U. S. Infantry at Fort B——.

Names.	Rank.	Company.	Number of shots.	Distance.	Length of string.
J—— D——	Corporal.	A	10 Rounds.	200 Yards.	20.1
C—— M——	Private.	B			19.5
F—— K——	Sergeant.	C			15.2
J—— L——	Private.	D			21.7

I certify that the above is a correct report of the trial for the best shot in the Companies of the 1st Regiment U. S. Infantry, stationed at Fort B——, June 1, 1863.

J—— M——,
Captain 1st U. S. Infantry.
Commanding Post.

In other cases the Board may render their opinion upon their own knowledge.

Return of Killed, Wounded, and Missing.

251. After an engagement, the company commander should make out as soon as possible a report according to Form 21. This report should be sent in to the adjutant at once by each company commander, in order that the result may be ascertained as soon as possible throughout the entire command.

252. The report should give the name, rank, and character of the wounds of the killed and wounded, and the circumstances of the missing as far as known.

Report of Target Practice.

253. When a company is detached from the regiment, it is necessary for the company commander to make reports of the firings to regimental head-quarters, for the information of the regimental commander, at such times and as often as the commanding officer may require. When the commanding officer fails to call for these reports, the company commander should, at least, at the termination of the annual exercises, send a consolidated report, showing the result of the exercises, and who the successful marksman is for the company prize, according to Form 1, Target Tactics.

254. A report is also necessary from the commanding officer of the post of the best shot in all the companies of the same regiment at the post, according to Form 22. (See Target Practice, page 39.)

Charges and Specifications.

255. It is recommended to company commanders to resort, in all cases where it is practicable, to legal punishments in the government of soldiers. To this end it is necessary, when offences are committed, to make out charges and specifications against the offenders, and send them in to the commanding officer of the regiment or post, who will either detail a field-officer, or, if necessary, make application for a general court-martial for their trial and punishment.

256. The following circular, from General McClellan's head-quarters whilst in command of the Army of the Potomac, gives nearly all the information necessary for drawing up charges and specifications :—

HEAD-QUARTERS, ARMY OF THE POTOMAC, ⎱
WASHINGTON, Aug. 28, 1861.   ⎰

In order that a well-established and generally understood method should obtain in the framing of charges exhibited against all persons amenable to military law, and those irregularities obviated which have heretofore retarded trial and impeded the administration of justice, the following form for the drawing up of charges and the accompanying instructions will be strictly adhered to throughout this army.

1. All charges preferred against an officer or soldier, and the circumstances upon which the charges are founded, must be previously examined by the officer immediately commanding the force wherein they originate. Commanders will be guided, as to the necessity of bringing an accused person to trial, *by the character of the individual,* his *conduct,* the nature and degree of the offence, its *prevalence* at the time in the *regiment,* and also by the *probabilities of conviction.*

2. Instead of embodying the crime to be tried in a single statement or indictment, called the charge, it is usual to try it in the form of a *charge* and *specification.*

3. *Of the Charge.*—Two methods are recognized in the drawing up of the charge,—the one by expressing the crime,

as the "*violation of such or such an article of war;*" the other by couching it in terms, either expressed or implied, of that article which contains the "*corpus delicti,*" ex. gr., "Disobedience of orders," "Desertion," "Mutiny," "Absence without leave," "Sleeping on post," "Drunkenness on duty," &c. &c. &c. Of these two modes the latter is deemed preferable, *and will, as far as practicable, be strictly followed.*

Apart from the vagueness which may attend the other, from the number of distinct crimes sometimes contained in the same "Article of War," it places the crime more closely before the accused, and, where enlisted men are concerned, their comprehension more easily discerns and more clearly understands the matter upon which they are arraigned.

4. *Of the Specification.*—The specification should be a simple statement of the *particular facts* and *circumstances* which make up the crime, with the adjuncts of *person, time,* and *place.* Where *language* or *gesture* are elements of the crime, they should both be stated with as much accuracy as possible. Besides the natural relation which should exist between a charge and its specification, the latter should be, as far as practicable, *certain* as to *time* and *place,* and a *concise, clear, and unequivocal* statement of the facts, devoid of all irrelevant matter; and under *no circumstances* should *two or more separate* and *distinct* crimes be *contained* in *the same specification.*

5. When it is desirable to induce the highest degree of punishment contemplated by any particular Article of War, it is *indispensably necessary* that the crime should be couched in the express terms of that Article. Thus, any prisoner accused of a violation of the "9th Article of War" (disobedience of orders), &c. &c., in order to convict him and award the punishment of death it would become necessary

to *specify* that he struck, lifted up, &c. &c. or disobeyed the lawful command of, his *superior officer whilst in the execution of his office :* were these terms omitted, and though it appeared in evidence that the ʼprisonerʼs offence was clearly the one contemplated by the 9th Article, he could not be visited by the severest punishment it may award.

6. The Christian and surname of the accused and witnesses, with their companies and regiments, must be stated; and the same care should be followed regarding all persons connected with the offence to be tried and named in the specification.

7. The charges and specifications having been drawn in accordance with the above instructions, they will be signed by the officer preferring them, or by the judge-advocate of the court before which they are presented for trial.

8. All charges of a *minor* character, referring to cases not capital, and which may be tried under the 99th Article of War, as "conduct prejudicial to good order and military discipline," will be brought before a regimental or garrison court-martial.

9. Regimental commanders will give the strictest attention to all charges submitted to them for their approval; and it is particularly enjoined upon the commanders of divisions and brigades to cause all charges to undergo a careful examination as regards their *form, legality,* and *neatness,* before being forwarded to these head-quarters.

By command of Major-General McClellan.

SETH WILLIAMS, *Assistant Adjutant-General.*

FORM OF STATING THE CHARGE.

1. Conduct prejudicial to good order and military discipline.

2. "Contempt and disrespect towards his commanding officer."

3. Mutiny.

4. "Striking, drawing, or lifting up a weapon upon his superior officer."

5. "Disobedience of orders."

6 "Signing a false certificate."

7. Desertion.

8. Absence without leave.

9. Drunkenness on duty.

10. Sleeping upon post.

11. Misbehavior or cowardice before the enemy.

12. "Disclosing the countersign."

13. "Forcing a safeguard."

14. Giving aid and comfort to the enemy.

15. Holding corespondence with, and giving intelligence to, the enemy.

16. "Breach of arrest."

17. Conduct unbecoming an officer and a gentleman.

FORM 1.—*Charge and Specification preferred against Private John A. Smith, Company A, 1st Regiment U. S. Infantry.*

CHARGE—*"Absence without leave."*

Specification—That Private John A. Smith, Company "A," 1st Regiment U. S. Infantry, was absent from his company, without permission from proper authority, between the hours of 12 and reveillé, on the night of the 15th and 16th of August, 1861.

This at Camp Scott, N. Y.

[Signed] ——— ———,

Officer preferring charge, or Judge-Advocate

Witnesses,

Private James Murray, Company "A," 1st Infantry.

"   T. O'Brien,   "   "   "

11

FORM 2.—*Charge and Specification preferred against Private John A. Smith, Company "A," 1st Regiment U. S. Infantry.*

CHARGE—*"Desertion."*

Specification—That Private John A. Smith, Company "A," 1st Regiment Infantry, having been. duly enlisted into the service of the United States, did desert the same on or about the 16th of August, 1861.

[Signed] ——— ———,
Officer preferring charge, or *Judge-Advocate.*

Witnesses,
Private J. Murray, Company "A," 1st Infantry.
"   Pat Riley,   "   "   "
"   Jas. G. Brien, "   "   "

FORM 3.—*Charge and Specification preferred against Private John A. Smith, Company "A," 1st Regiment U. S. Infantry.*

CHARGE—*"Sleeping on Post."*

Specification—That Private John A. Smith, Company "A," 1st Regiment Infantry, having been duly posted as a sentinel, was found asleep on his post, between the hours of 12 M. and 1 A.M., when visited by the corporal of the guard.

This at Camp Scott, N. Y., on or about the 15th August, 1861.

[Signed] ——— ———,
Officer preferring charge, or *Judge-Advocate.*

Witnesses,
Corporal John Jones, Company "A," 1st Infantry.
Private S. Murphy,   "   "   "
    &c.   &c.   &c.   &c.

FORM 4.—*Charge and Specification preferred against Private John A. Smith, Company "A," 1st Regiment U. S. Infantry.*

CHARGE—*"Drunkenness on Duty."*

Specification—That Private John A. Smith, Company "A," 1st Regiment Infantry, was drunk at evening parade on the evening of the 15th of August, 1861.

This at Camp Scott, on or about the 15th of August, 1861.

[Signed] ——— ———,
Officer preferring charge, or *Judge-Advocate.*

Witnesses,

Corporal John Jones, 1st Regiment Infantry.
Private S. Murphy, " " "

FORM 5.—*Charges and Specifications preferred against Private John A. Smith, Company "A," 1st Regiment U. S. Infantry.*

CHARGE 1—*"Disobedience of Orders."*

Specification—That Private John A. Smith, Company "A," 1st Regiment Infantry, having been ordered by his superior officer, Capt. A. Brown, 1st Regiment Infantry, to fall into the ranks of his company (the said Capt. Brown being in the execution of his office), did wilfully refuse to obey said order, and did reply in words or figures to wit: "I'll be d——d if I will fall in," or words to that effect.

This at Camp Scott, N. Y., on or about the 15th of August, 1861.

CHARGE 2—*"Disrespect to his Superior Officer."*

Specification—That the aforesaid private John A. Smith, Company "A," 1st Regiment Infantry, having been ordered by his superior officer, Capt. A. Brown, 1st Infantry, to fall into the ranks of his company, did reply to the said Capt. Brown in a disrespectful manner, in words and figures to wit: "I'll be d——d if I fall in," or words to that effect.

This at Camp Scott, N. Y., on or about the 15th of August, 1861.

Capt. —— ——, 1st Regiment Infantry,
or Judge-Advocate of the Court.

Witnesses,

Private John Murray, Company "A," 1st Reg't. Infantry.

"   Patrick Riley,   "   "   "   "

"   James O'Brien,   "   "   "   "

(Note to Par. 8 in the circular.) By sec. 7 of the Act of July 17, 1862, a field-officer is authorized to act in the cases alluded to in this paragraph. In the absence of a field-officer the commanding officer cannot do otherwise than detail a regimental or garrison court-martial.

"SEC. 7. *And be it further enacted*, That hereafter all offenders in the army, charged with offences now punishable by a regimental or garrison court-martial, shall be brought before a field-officer of his regiment, who shall be detailed for that purpose, and who shall hear and determine the offence, and order the punishment that shall be inflicted; and shall also make a record of his proceedings, and submit the same to the brigade commander, who upon the approval of the proceedings of such field-officer shall order the same to be executed: *Provided*, That the punishment in such case be limited to that authorized to be inflicted by a regimental or garrison court-martial; and *provided*, further, That in the event of there being no brigade commander, the proceedings as aforesaid shall be submitted for approval to the commanding officer of the post."

The following section of the Act of March 3, 1863, creates additional offences, over which military courts have jurisdiction in time of war:—

"Sec. 30. *And be it further enacted*, That in time of war insurrection or rebellion, murder, assault and battery, with an intent to kill, manslaughter, mayhem, wounding by shooting or stabbing, with an intent to commit murder, robbery, arson, burglary, rape, assault and battery, with an intent to commit rape and larceny, shall be punishable by the sentence of a general court-martial or military commission when committed by persons who are in the military service of the United States and subject to the Articles of War; and the punishment for such offences shall never be less than those inflicted by the laws of the State, Territory, or District in which they may have been committed."

Letters of Transmittal.

257. All returns, reports, and other papers to higher authority should be accompanied by a letter of transmittal, stating the character of the papers transmitted, and giving any explanations that may be necessary.

258. A letter to higher authority should be signed by the officer himself who sends it, and not by some other officer under his direction, and should be directed to the adjutant, if intended for the colonel or post commander, and to the Assistant Adjutant-General, if intended for brigade or division head-quarters or the head-quarters of the army. Superiors address their communications through their staff officers to the commanders direct that are subordinate to them, and not to their adjutants or staff officers.

259. When possible, all letters of transmittal and other papers should be folded the size of ordinary letter-paper folded in three folds. The dimensions are three and a half inches wide and eight inches long. Letters of transmittal and other official letters, when endorsed, should give, first the place and date, then the name and rank of the

11*

writer and his regiment, and, finally, a brief summary of the contents, as follows, viz. :—

CAMP CHASE, OHIO, January 3, 1863.

Capt. A—— B——,
    2d U. S. Cavalry,
        Commanding Company A.

Letter transmitting Quarterly Return of Clothing, Camp and Garrison Equipage, for the fourth quarter, 1862.

---

260. Some departments do not require endorsements, preferring to do it themselves in their own offices, to insure their correctness. The above, however, is the correct endorsement. It should be made on the first or top fold.

261. The following is the usual form of all letters. The address may be either at the top or bottom of the letter, viz. :—

CAMP CHASE, OHIO, January 5, 1863.

SIR :—

I have the honor to enclose herewith Quarterly Returns of Clothing, Camp and Garrison Equipage, in duplicate for the 4th quarter of 1862, with one complete set of vouchers.

      Very respectfully, your ob't serv't, .
          A—— B——,
            Capt. 2d U. S. Cavalry, Comm'dg Co. A.
GEN. M. C. MEIGS,
    Quartermaster-General U. S. A.,
        Washington, D. C.

262. All official letters should be dated, signed, and addressed in the foregoing form. Each letter should, as far as possible, relate to one subject, and matters relating to and affecting different departments of the army should in no case be put in the same letter, but each should be made the subject of a separate communication.

263. Reports of service performed, or transactions of the company, are also made the subject of a letter in the foregoing form, as the report of a skirmish or reconnoissance.

Complaints and Applications of Soldiers.

264. Soldiers are frequently totally incapable of putting their complaints or applications in writing, for the want of education; and, as it is often necessary that they should be a matter of record, especially where they are to be referred to higher authority, the company clerk should be required to put them in proper form. An example is given below.

CAMP CHASE, OHIO, January 21, 1863.

SIR :—

I have the honor to call the attention of the commanding officer of the regiment to certain acts of ill treatment that I have received at the hands of Lieut. J—— C——, 2d Lieut. 1st U. S. Infantry. (Here give the circumstances of the ill treatment in detail.)

Respectfully submitted for the action of the commanding officer of the regiment.

Very respectfully, your ob't serv't,

C—— D——,

Private Co. A. 1st U. S. Infantry.

Lieut. L—— M——,

1st Lieutenant and Adjutant.

1st U. S. Infantry.

Approved and forwarded,

A—— B——,

Capt. Company A, 1st U. S. Infantry.

265. The 35th Article of War requires the commanding officer of the regiment to order a regimental court-martial in cases like the foregoing, from whose action either party may appeal to a general court-martial, whose action will be final.

266. The following is the form of an application to be transferred from one company to another in the same regiment, viz. :—

CAMP CHASE, OHIO, January 20, 1863.

SIR :—

I have the honor to apply to be transferred from Company "A" to Company B, for the following reasons. (Here state the reasons for making the application.)

Very respectfully, your obedient servant,

A—— G——,

Private Co. A, 1st U. S. Infantry.

Lieut. L—— M——,

1st *Lieut. and Adjutant,*

1st *U. S. Infantry.*

Approved,

J—— P——,

Capt. Co. A, 1st U. S. Infantry.

J—— J——,

Capt. Co. B, 1st U. S. Infantry.

In cases of applications for a transfer in the same regiment, it is necessary to have the approval of the commanding officers of the companies to which the soldier belongs and to which he desires to transfer, before submitting to the regimental commander. Where the application is to be transferred from one regiment to another, in addition, it is necessary to have the approval of the commanding officers of both regiments, before it is sent up to the commanding general. In the same regiment, but to different posts, it is necessary to have the order of the department commander.

Leave of Absence.

267. Applications for leave of absence must be made in writing, when desired by officers, and must be endorsed either *Approved* or *Disapproved* by the intermediate commanders, as in the case of a lieutenant by his captain, the commanding officer of the regiment, brigade, and division, and corps or army commander, until it arrives at the authority competent to grant the desired leave. It may be addressed to the first or last commander, always remembering that the adjutant and adjutant-generals are the representatives of the commanders in such cases.

268. The length of, and authority to grant, leaves of absence are regulated in Orders, in time of war. In time of peace the commander of a post may grant leaves of absence not to exceed seven days at one time in the same month; the commander of a department, not to exceed sixty days; the commanding general of the army, not to exceed four months; and for any longer period the application must be submitted to the Secretary of War.

269. An application including in it permission to apply for an extension, the term of the extension should be stated. Applications for a leave of absence on account of sickness should be accompanied by a surgeon's certificate. (See Form, par. 185, Reg.)

270. The form of an application is that of an ordinary official letter, viz. :—

CAMP CHASE, OHIO, January 15, 1863.

SIR :—

I have the honor to request a leave of absence for sixty days, for the purpose (here state the locality he desires to visit, and the reasons therefor), with permission to apply for an extension of two months.

I have not received a leave of absence longer than seven

days since (here state the time the officer has been without a leave of absence).

Captain J—— L—— and Lieutenant C—— D—— are both serving with the company to which I am attached, and my services can readily be spared for the period desired.

Very respectfully, your obedient servant,

L—— M——,

Major J—— C——,   1st *Lieut.* 2d *U. S. Cavalry.*
Assistant Adjutant-General,
Department of P——.

271. Leaves of absence are rarely refused where there are two other officers serving with the company to which the officer belongs, in time of peace. It is, therefore, commendable in officers to serve with their proper companies as much as possible, in order that the indulgence of leaves of absence be shared alike.

The following law has recently been passed (Act March 3, 1863):—

"SEC. 31. *And be it further enacted,* That any officer absent from duty with leave, except for sickness or wounds, shall, during his absence, receive half of the pay and allowances prescribed by law, and no more; and any officer absent without leave shall, in addition to the penalties prescribed by law or a court-martial, forfeit all pay or allowances during such absence."

Resignations.

272. The same rules apply for forwarding resignations as for applications for leaves of absence. The resignation is not final until it has been accepted.

273. Resignations are usually sent to the Adjutant-General of the army, except in case where an officer resigns on account of ill health, and the resignation is accompanied by a surgeon's certificate to the effect that the officer is rendered unfit for duty.

274. Resignations tendered under charges should be accompanied by a copy of the charges, or a statement of the case in the absence of written charges.

275. An officer is not permitted to resign until he has rendered all his accounts, and has turned over or is prepared to turn over all the Government property for which he is accountable. When the date at which the resignation is to take effect is not stated, it is limited to thirty days from the date of the order.

276. To enable an officer to leave his command at once on leave of absence, the resignation must be unconditional and immediate, and the officer's address should be given, where the acceptance may be sent to him.

The following form is customary, viz.:—

CAMP CHASE, OHIO, Jan. 17, 1863.

SIR:—I have the honor herewith to tender my resignation as 2d lieutenant in the 1st Regiment U.S. Infantry, on account of (here the officer will state any reason which he may think proper).

I certify, on honor, that I am not indebted to the United States on any account whatever, and that I am not responsible for any Government property, except what I am prepared to turn over to the proper officer on the acceptance of my resignation, and that I was last paid by Major D—— C—— to include the 31st day of December, 1862.

Very respectfully, your obedient servant,
C—— M——,

Gen. L—— T——,   2d Lieut. 1st U. S. Infantry.
Adj't-Gen. U. S. A., Washington, D. C.

Enlistments and Re-enlistments.

277. The company may be kept up to the maximum

strength by enlistments and re-enlistments in the company; but the maximum cannot be exceeded. In infantry the maximum is 98 enlisted men, in cavalry 100, and in a battery of artillery 147. The following are the organizations as allowed by existing laws (see Gen. Orders No. 126, 1862, and Act March 3, 1863, sec. 37) :—

Company of Infantry.

1 Captain,	4 Sergeants,
1 First Lieutenant,	8 Corporals,
1 Second Lieutenant,	2 Musicians,
1 First Sergeant,	1 Wagoner,

And $\begin{cases} 64 \text{ Privates—minimum.} \\ 82 \text{ Privates—maximum.} \end{cases}$

Company or Troop of Cavalry.

1 Captain,	1 First Sergeant,
1 First Lieutenant,	1 Quartermaster Sergeant,
1 Second Lieutenant,	1 Commissary Sergeant,
5 Sergeants,	2 Farriers or Blacksmiths,
8 Corporals,	1 Saddler,
2 Trumpeters,	1 Wagoner, and

78 Privates.

Battery of Artillery.

1 Captain,	4 Sergeants,
1 First Lieutenant,	8 Corporals,
1 Second Lieutenant,	2 Musicians,
1 First Sergeant,	2 Artificers,
1 Quartermaster Sergeant,	1 Wagoner, and

122 Privates.

To the above organization of a battery one first and one second lieutenant, two sergeants, and four corporals may be added, at the President's discretion. The authority for this addition is generally given for a full battery of six pieces.

278. The commanding officer of each regiment generally details an officer—usually a lieutenant—to attend to the recruiting in the regiment. He can be supplied with funds for the payment of bounty through the commanding officer of the regiment. All soldiers enlisting or re-enlisting in the regiment should be referred to him.

279. One hundred dollars bounty is now allowed to each soldier who serves two years or more. A soldier who enlists for this or a longer period is entitled to an advance of $25 of this bounty (Act July 22, 1861, sec. 5), and a premium of $2. (Act June 21, 1862.)

280. Volunteers, not to exceed ten from any one company, have been permitted to enlist in the regular service to serve out the unexpired portion of their enlistments, or for a full term. (See Gen. Orders Nos. 154 and 162; the Act of March 8, 1863, sec. 36, rescinds these orders; but those who enlisted prior to Feb. 10, 1863, are entitled to the benefits of them. If they have received the above advance and premium, they are not entitled to it again, but receive the balance at the expiration of their term of service.

281. In the absence of a recruiting officer, company commanders may enlist soldiers in their own companies, or re-enlist men whose terms of service have expired; and in the absence of funds to pay the bounty and premium, it may be paid by the paymaster on the first pay-roll.

282. Triplicate enlistment papers are made out by the officer enlisting the soldier. One is sent to the Adjutant-General at Washington, the other to the adjutant of the regiment, and the last is retained by the officer. Blanks for the purpose are obtained like any other blanks furnished by the Adjutant-General's Department.

283. The re-enlistment of a soldier is precisely similar to an enlistment, except that re-enlistment blanks are used, which show the length and nature of former services.

Volunteers and Militia.

284. Volunteers and militia are generally called out under some law authorizing the call, or in a great emergency by the Governor of a State or Territory, trusting to the approval of Congress and a reimbursement of the expenses incurred.

285. When volunteers are mustered into the service of the United States they are subject to the same laws and regulations as the troops of the Regular Army, except so far as the special law under which they are called out may modify them.

286. "No officer or enlisted man of volunteers is properly in the service of the United States, or authorized to receive pay, until mustered in by the proper officer; and no officer is properly out of service until discharged in orders, or mustered out by the proper officer." (G. O. No. 48, W. D. 1863.)

287. Mustering officers are generally appointed from the Regular Army to muster volunteers into service; they receive their instructions from the War Department. They are stationed at convenient points, or appointed for specific commands, and to these officers volunteers must apply to be mustered in or out of service.

288. Volunteer officers must produce a commission or appointment from the Governor of the State before they can be mustered in, and for every new grade to which they attain they must be mustered out of the old and mustered into the new. Enlisted men must be discharged before they can be mustered in as commissioned officers. Musters cannot be dated back, except by authority from the Adjutant-General, and payment commences on the date of muster. No officer can be mustered into service unless a vacancy exists. Officers organizing troops must have men enough to correspond to their rank; thus, when one-half the company has been mustered into service, the first lieutenant can be mustered in, and when the company is complete, the captain and second lieutenant can be mustered in.

Field and staff officers of regiments can be mustered into service upon the completion of the organization of regiment or companies, as follows :—

INFANTRY.

Colonel and chaplain—entire regiment.
Lieutenant-colonel—four companies.
Major—six companies.

CAVALRY.

Colonel and chaplain—entire regiment.
Lieutenant-colonel—six companies.
Majors—one for every four companies.

ARTILLERY.

Colonel and chaplain—entire regiment.
Lieutenant-colonel—six companies.
Majors—one for every four batteries.

289. In organizing a volunteer company the enlisted men can be mustered in from date of enrollment. The captain causes a complete roll to be made; blanks for the purpose can be obtained from the mustering officer. The names are entered on the roll in the order of rank in every grade, and the privates alphabetically. It is the mustering officer's duty to reject all officers or soldiers physically disqualified. (For more complete information on the subject of mustering in and out of service, see *Instructions for making Muster Rolls, &c.*, and Gen. Order No. 48, 1863, issued by the War Department.)

TABULAR LIST OF ROLLS, RETURNS, AND REPORTS

DESIGNATION.	WHEN TO BE MADE.
To Adjutant-General.	
Muster Roll of Company*.......................................	Every two months............
Inventory of Effects of Deceased Soldiers...............	Immediately....................
Final Statements of Deceased Soldiers....................	Do. ....
To the Quartermaster-General.	
Duplicate Returns of Clothing, Camp and Garrison Equipage, and Quartermaster's Property — one with and one without vouchers...........................	End of every quarter.......
To the Chief of Ordnance.	
Returns of Ordnance and Ordnance Stores..............	End of every quarter.......
Certificate of Inventory on Return of Ordnance and Ordnance Stores..	Yearly—in June..............
Report of Damaged Arms....................................	End of every two months.
To Regimental Head-Quarters.	
Descriptive List of Men Joining..............................	End of every quarter.......
Return of Deceased Soldiers..................................	Do. ....
Return of the Company..	End of every month.........
Transcript of Orders making temporary appointments of Non-commissioned Officers, or reducing Non-commissioned Officers, at Posts not Regimental Head-Quarters...	Immediately................,.........
Inventory of Effects of Deceased Soldiers...............	Do. ....
Final Statements of Deceased Soldiers....................	Do. ....
Return of Blanks..	End of every quarter.......
To Post Adjutant.	
Morning Report of Company..................................	Each morning,..................
Monthly Return of Company (to be returned to Company Commander for file)...............................	End of every month.........
Return of Company Fund, with Company Council Book	End of every four months.

The same Returns as above are to be made by Officers commanding Bands or Company Officers, when on Regimental Recruiting Service, make to the TENDENT (*Regimental Commander*), the same Reports and Returns as rendered by When soldiers die possessed of no effects, the fact will be so stated both upon Returns of Deceased Soldiers will be forwarded even in cases where no deaths properly headed and signed, with a black or red ink line drawn obliquely across

* *Three* MUSTER AND PAY ROLLS are made out at the same time—*two* for the Pay-same time forwarded to the Adjutant-General.

REQUIRED FROM COMPANY COMMANDERS.

WHEN TO BE SENT.	BY WHOM.	PARAGRAPH OF REGULATIONS, 1861.
Within three days thereafter.....	Mustering Officer.......	333 and 334.
Immediately	Company Commander.	152.
Do.	Do.	152.
Within twenty days thereafter...	.............Do	1169, 1171, and 1173.
Within twenty days thereafter...	.............Do	1421.
Do. do. ...	.............Do	1425.
First day of subsequent month...	.............Do	1395.
First day of subsequent month...	.............Do	88 (4th clause).
Do. do. ...	.............Do	463.
Do. do. ...	.............Do	458.
Immediately	.............Do	74 and 79.
Do.	.............Do	152.
Do.	.............Do	152.
As soon as completed................	.............Do	Gen. O., No. 13, 1862.
Before eight o'clock A.M............	.............Do	236.
First of subsequent month.........	.............Do	458.
Do. do.	.............Do	206.

small detachments of troops.
ADJUTANT-GENERAL, QUARTERMASTER-GENERAL, CHIEF OF ORDNANCE, and SUPERIN-
Officers on the General Recruiting Service. (Par. 985, Regulations of 1861.)
the Inventory and the Final Statement. (Par. 152, Regulations, 1861.)
have occurred during the quarter. In such cases, blank forms will be forwarded,
the body of the Return from left to right.

master and *one* to be retained with the Company. *One* MUSTER ROLL is at the

INDEX.